高校土木工程专业规划教材

土木工程专业英语（第二版）

Specialty English of Civil Engineering (2nd Edition)

崔春义 张 鹏 刘海龙 赵九野 主 编

裴华富 主 审

中国建筑工业出版社

图书在版编目（CIP）数据

土木工程专业英语/崔春义等主编．—2版．—北京：中国建筑工业出版社，2019.10
高校土木工程专业规划教材
ISBN 978-7-112-24197-2

Ⅰ．①土… Ⅱ．①崔… Ⅲ．①土木工程—英语—高等学校—教材 Ⅳ．①TU

中国版本图书馆CIP数据核字（2019）第202999号

全书课文以土木工程为主线，全面系统地介绍了土木工程及其所属分支有关学科的基本内容，如建筑、桥梁、道路、交通、材料、管理等。书中还给出了土木工程直接相关的数学和力学（理论力学、材料力学和结构力学）的常用英语表达。全书选文精泛结合，简繁有度，并配有部分参考译文，有利于全面培养学生的实践能力和自学能力。

本书主要作为普通高等学校土木工程专业、交通土建专业本科生和研究生教育使用的专业英语教材或参考书，也可作为土木科技、设计施工技术人员学习专业英语的自学用书，亦可作为高职高专、（成人）专升本及电大、自考等继续教育学生的学习教材。

* * *

责任编辑：辛海丽　王　梅
责任校对：芦欣甜

高校土木工程专业规划教材
土木工程专业英语
（第二版）
Specialty English of Civil Engineering (2nd Edition)
崔春义　张　鹏　刘海龙　赵九野　主　编
裴华富　主　审

*

中国建筑工业出版社出版、发行（北京海淀三里河路9号）
各地新华书店、建筑书店经销
北京科地亚盟排版公司制版
天津翔远印刷有限公司印刷

*

开本：787×1092毫米　1/16　印张：13½　字数：327千字
2019年12月第二版　2019年12月第三次印刷
定价：39.00元
ISBN 978-7-112-24197-2
（34692）

版权所有　翻印必究
如有印装质量问题，可寄本社退换
（邮政编码100037）

第二版前言

学好专业外语是掌握学科发展动态、获取专业信息、参加国际学术交流的基本前提。本教材紧密结合土木工程专业的"新工科"课程视角，在第一版内容基础上，通过精心选择和重新校对，并增加相关英语专业文献素材和典型参考译文后编写而成。其选文精泛结合，简繁有度，对常用词汇进行了重新梳理总结。全书课文以土木工程为主线，系统地介绍了土木工程及所属或相关专业的基本内容，内容具体涵盖建筑工程、道路工程、交通工程、隧道工程、桥梁工程、建筑工程管理等诸多专业方向。本书共分13章，含34篇课文，最后附录中还给出了土木工程专业工程和常用相关力学词汇汉英表达。本书可作为高校土木工程领域本科生、研究生"专业英语"教材或参考书，也可作为成人专升本及土木科技与工程技术人员学习专业英语的自学用书。

本书第二版修订工作由大连海事大学土木工程系崔春义教授主编并统稿，由大连理工大学裴华富教授主审。全书具体修订工作按章节分工如下：第1、3、6、10~13章由张鹏博士和崔春义教授编写修订，第2、7、8章由刘海龙博士编写修订，第4、5、9章由赵九野博士编写修订。

参与本书图片和文字编辑工作的还有大连海事大学博士研究生孟坤、梁志孟、辛宇、王本龙、于春阳以及硕士研究生厉超吉、许民泽、姚怡亦、张新程、阮文宁等，在此表示感谢。

本书的出版得到了采薇君华公司教育部产学研合作协同育人项目（201801171013）、晨曦科技股份有限公司教育部产学研合作协同育人项目（201802132058）和大连海事大学教学改革项目（2018Y36）的资助。

本次书稿第二版编写和修订工作是利用业余时间完成，由于时间仓促，加之编者水平有限，书中的疏误之处在所难免，恳请同行专家及广大读者提出宝贵批评意见，以利修改。

编　者

2019年7月20日　于大连海事大学　路桥楼

第一版前言

作为新时代的土木工程师,除了具备坚实的专业知识还应拥有良好的科技外语交流能力。另外,学好专业外语也是掌握学科发展动态、获取专业信息、参加国际学术交流的基本前提。

本教材紧密结合土木工程专业,通过精心选择相关英语专业文献素材编写而成,选文精泛结合,简繁有度。全书课文以土木工程为主线,系统地介绍了土木工程及其所属分支有关学科的基本内容,内容涵盖建筑工程、道路工程、交通工程、隧道工程、桥梁工程、建筑工程管理等诸多专业方向。本书可作为高校土木工程领域本科生、研究生"专业英语"教材或参考书,也可作为成人专升本及土木科技与工程技术人员学习专业英语的自学用书。

本书由大连海事大学土木工程系崔春义教授主编并统稿,由北京工业大学何浩翔教授主审。本书共分12章,含32篇课文,最后附录中还给出了土木工程专业工程和常用相关力学词汇汉英表达。参与本书图片和文字编辑工作的还有大连海事大学博士研究生孟坤、张石平,硕士研究生朱江山、王春乐、赵飞翔、张叶林、赵会杰等,在此表示感谢。

本书是利用业余时间编写,由于时间仓促,加之编者水平有限,书中的疏误之处在所难免,恳请同行专家及广大读者提出宝贵批评意见,以利修改。

编　者

2014年11月8日　于大连海事大学　路桥楼

目 录

LESSON 1　INTRODUCTION TO CIVIL ENGINEERING ·················· 1
　Text A　Scope of Different Fields of Civil Engineering ························· 1
　参考译文：土木工程相关专业与学科领域 ································· 4
　Text B　Infrastructure ··· 6
LESSON 2　SOIL MECHANICS ··· 8
　Text A　The Origin of Soils ··· 8
　参考译文：土体的来源 ·· 12
　Text B　Vertical Stresses in a Layer ··· 13
LESSON 3　ENGINEERING MECHANICS ·· 19
　Text A　Types of Beams, Loads, and Reactions ······························ 19
　参考译文：梁、荷载及反力类别 ··· 22
　Text B　Types of Loads and Reactions ·· 24
　Text C　Changes in Lengths under Nonuniform Conditions ·············· 30
　Text D　The Determination of Changes in Lengths of Nonuniform Bars ······· 33
LESSON 4　CONSTRUCTION MATERIALS ···································· 37
　Text A　Structural Steels ··· 37
　参考译文：结构钢材 ·· 42
　Text B　Asphalt Binders and Asphalt Mixtures ······························ 44
　Text C　Portland Cement Concretes ·· 48
LESSON 5　DESIGN OF CONCRETE STRUCTURE ························ 54
　Text A　Design of Concrete Beams ·· 54
　参考译文：混凝土梁的设计 ··· 58
　Text B　Design of Prestressed Concrete ··· 61
LESSON 6　DESIGN OF STEEL STRUCTURE ································ 67
　Text A　Introduction to Limit State Design ···································· 67
　参考译文：极限状态设计简介 ·· 69
　Text B　Analysis Procedures and Design Philosophy ························ 71
LESSON 7　FOUNDATION ENGINEERING ···································· 74
　Text A　Types of Failure in Soil at Ultimate Load ·························· 74
　参考译文：极限荷载作用下土体破坏模式 ······································· 81
　Text B　Piled Foundation Choice ·· 83
　Text C　Sustainability and Environment ·· 87

LESSON 8　TUNNEL ENGINEERING ... 89
    Text A　Tunnel Engineering ... 89
    参考译文：隧道工程 ... 94
    Text B　Tunnel Linings ... 96

LESSON 9　HIGHWAY ENGINEERING ... 104
    Text A　Highway Alignments ... 104
    参考译文：公路线形 ... 109
    Text B　Road Surfaces ... 112

LESSON 10　BRIDGE ENGINEERING ... 118
    Text A（1）　Steel Bridges ... 118
    参考译文：桥梁工程 ... 122
    Text A（2）　Truss Bridges and Suspension Bridges ... 125
    Text A（3）　Cable-Stayed Bridges and Arch Bridges ... 129
    Text B（1）　Concrete Bridges ... 134
    Text B（2）　Concrete Bridge Piers and Abutments ... 138

LESSON 11　BUILDING ENGINEERING ... 141
    Text A　Structural Systems ... 141
    参考译文：结构系统 ... 145
    Text B　Lateral-Force Bracing ... 147

LESSON 12　CONSTRUCTION MANAGEMENT ... 151
    Text A　Tasks of Construction Management ... 151
    参考译文：施工管理任务 ... 154
    Text B　Organization of Construction Firms ... 156
    Text C　Prime Contracts and Subcontracts ... 162

LESSON 13　BUILDING INFORMATION MODELING ... 167
    Text A　Basic Concepts of BIM ... 167
    参考译文：建筑信息模型（BIM）基本知识 ... 170
    Text B　PARAMETRIC MODELING OF BUILDINGS ... 173

附录　土木工程专业词汇汉英表达 ... 178
    附录A　结构工程与防灾工程 ... 178
    附录B　桥梁工程 ... 179
    附录C　岩土工程 ... 183
    附录D　结构动力学 ... 189
    附录E　理论力学 ... 192
    附录F　材料力学 ... 194
    附录G　结构力学 ... 197
    附录H　基本术语 ... 199

附录 I 常用符号 ·· 200
附录 J 数学知识 ·· 200
附录 K 图形名称 ·· 202
附录 L 常用数学表达式 ·· 203
附录 M 常用希腊字母 ·· 205
附录 N 读法实例 ·· 206
附录 O 数学问题求解的一般表示 ·· 207
附录 P 进制 ·· 207
参考文献 ·· 208

LESSON 1   INTRODUCTION TO CIVIL ENGINEERING

Text A   Scope of Different Fields of Civil Engineering

Civil engineering is the oldest **branch** of engineering which is growing right from the stone age civilization. American Society of Civil Engineering defines Civil Engineering as the profession in which a knowledge of the mathematical and physical sciences gained by study, experience and practice is applied with judgment to develop ways to utilize economically the materials and forces of nature for the progressive well-being of man.

In this chapter, scopes of different fields of civil engineering are discussed and the importance of developing infrastructure in the country is presented.

Civil Engineering may be divided into the following fields:

(ⅰ) **Surveying**
(ⅱ) **Building Materials**
(ⅲ) **Construction Technology**
(ⅳ) **Structural Engineering**
(ⅴ) **Geotechnical Engineering**
(ⅵ) **Hydraulics**
(ⅶ) **Water Resources and Irrigation Engineering**
(ⅷ) **Transportation Engineering**
(ⅸ) **Environmental Engineering**
(ⅹ) **Architecture and Town planning**

Scope of each one of these is discussed below.

(ⅰ) Surveying

Surveying is the science of map making. To start any development activity in an area, the relative positions of various objects in the horizontal and vertical directions are required. This is approved by surveying the area. Earlier, the conventional instruments like chain, tape and leveling instruments were used. In this electronic era, modern equipments like distance meters and total stations are used to get more accurate results easily. The modern technologies like photogrammetry and remote sensing have made surveying easier.

(ⅱ) Building Materials

Shelter is the basic need of civilization. To get good shelter, continuous efforts are going on right from the beginning of civilization. Stones, bricks, **timber, lime, cement,** sand, **jellies** and **tiles** are the traditional building materials. Use of steel, aluminum, glass, **glazed**

tiles, **plaster of paris**, paints and **varnishes** have improved the quality of buildings. The appropriate mixture of binding materials like lime and cement with sand is known as **mortar**. The mixture of cement, sand and jelly (crushed stones) with water is known as **concrete**. The use of concrete with steel bars placed in appropriate position has helped in building strong and durable tall structures. The composite material of concrete and steel is called **reinforced cement concrete** which is popularly known as RCC. A civil engineer must know the properties of all the building materials so that they can be used appropriately. Improved versions of many building materials appear in the market. A good civil engineer will make use of them at the earliest.

(iii) Construction Technology

Construction is the major activity of civil engineering which is continuously improving. As land cost is going up, there is demand fortall structures in urban areas while in rural areas need is for low cost constructions. One has to develop technology using locally available materials. In India, contribution of Central Building Research Institute (CBRI)-Roorkee and Gaziabad, several educational institutions throughout the country and Nirmithi Kendras in the technology development are noteworthy.

(iv) Structural Engineering

Load acting on a structure is ultimately transferred to ground. In doing so, various components of the structure are subjected to internal stresses. For example, in a building, load acting on a slab is transferred by slab to ground through beams, **columns** and footings. Assessing the internal stresses in the components of a structure is known as **structural analysis** and finding the suitable size of the structural component is known as **design of structure**. The structure to be analysed and designed may be of masonry, RCC or steel. Upto mid-1960's considerable improvements were seen in classical analyses. With the advent of computers numerical methods emerged and analyses and design packages were becoming popular. **Matrix Method of analysis** and **Finite Elements Analysis** have helped in the analysis of complex structures. A civil engineer has not only to give a safe structure but he has to give economical sections. To get economical section mathematical **optimization** techniques are used. Frequent earthquakes in the recent years have brought, importance of analysis of the structure for **earthquake forces**. Designing earthquake resistant structures is attracting lots of researches. All these aspects fall under structural engineering field.

(v) Geotechnical Engineering

Soil property changes from place to place. Even in the same place it may not be uniform at various depths. The soil property may vary from season to season due to variation in **moisture content**. The load from the structure is to be safely transferred to soil. For this, **safe bearing capacity** of the soil is to be properly assessed. This branch of study in Civil Engineering is called as Geotechnical Engineering.

Apart from finding safe bearing capacity for foundation of buildings, geotechnical engineering involves various studies required for the design of **pavements, tunnels, earthen**

dams, **canals** and **earth retaining structures**. It involves study of ground improvement techniques also.

(vi) Hydraulics

Water is an important need for all living beings. Study of mechanics of water and its flow characteristics is another important field in Civil Engineering and it is known as hydraulics.

(vii) Water Resources and Irrigation Engineering

Water is to be supplied to agriculture field and for drinking purposes. Hence suitable water resources are to be identified and water is to be stored. Identifying, planning and building water retaining structures like tanks and dams and carrying stored water to fields is known as water resources and irrigation engineering.

(viii) Transportation Engineering

Transportation facility is another important need. Providing good and economical roads is an important duty of civil engineers. It involves design of base courses, suitable surface finishes, cross drainage works, road intersections, **culverts**, bridges, tunnels etc. **Railway** is another important long-way transport facility. Design, construction and maintenance of railway lines, signal system are part of transportation engineering. There is need for airports and **harbors**. For proper planning of this transportation facility, traffic survey is to be carried out. Carrying out traffic survey, design, construction and maintenance of roads, bridges, railway, harbor and airports is known as transportation engineering.

(ix) Environmental Engineering

Proper distribution of water to rural areas, towns and cities and disposal of waste water and solid waste are another field of civil engineering. Industrialisation and increase in vehicular traffic are creating air pollution problems. Environmental engineering while tackling all these problems provides healthy environment to public.

(x) Architecture and Town Planning

Aesthetically good structures are required. Towns and cities are to be planned properly. This field of engineering has grown considerably and has become a course separate from Civil Engineering.

New Words and Expressions
[1] civil engineering 土木工程；
[2] branch [brɑːntʃ] *n*. 分支；
[3] surveying [sɜːˈveɪŋ] *n*. 测量学；
[4] building materials 建筑材料；
[5] construction technology 施工技术；
[6] structural engineering 结构工程；
[7] geotechnical engineering 岩土工程；
[8] hydraulics [haɪˈdrɔːlɪks] *n*. 水力学；
[9] water resources and irrigation engineering 水资源和灌溉工程；

［10］transportation engineering 运输工程；
［11］environmental engineering 环境工程；
［12］town planning 城镇规划；
［13］shelter ['ʃeltə(r)] n. 庇护所；
［14］timber ['tɪmbə(r)] n. 木材；
［15］lime [laɪm] n. 石灰；
［16］cement [sɪ'ment] n. 水泥；
［17］jellies ['dʒelɪz] n. 凝胶剂；
［18］tiles [taɪlz] n. 瓷砖；
［19］glazed tiles 釉面砖；
［20］plaster of paris 熟石膏；
［21］varnish ['vaːnɪʃ] n. 清漆；
［22］mortar ['mɔːtə(r)] n. 砂浆；
［23］concrete ['kɒnkriːt] n. 混凝土；
［24］reinforced cement concrete 钢筋水泥混凝土；
［25］column ['kɒləm] n. 柱；
［26］structural analysis 结构分析；
［27］design of structure 结构设计；
［28］matrix method of analysis 矩阵的分析方法；
［29］finite elements analysis 有限元分析；
［30］optimization [ˌɒptɪmaɪ'zeʃən] n. 优化；
［31］earthquake forces 地震作用；
［32］soil property 土壤性质；
［33］moisture content 水分含量；
［34］safe bearing capacity 安全承载能力；
［35］pavements ['peɪvmənts] n. 路面；
［36］tunnel ['tʌnl] n. 隧道；
［37］earthen dams 堤坝；
［38］canal [kə'næl] n. 运河；
［39］earth retaining structures 挡土结构；
［40］culvert ['kʌlvət] n. 涵洞；
［41］railway ['reɪlweɪ] n. 铁路；
［42］harbor ['hɑːbə] n. 港口。

参考译文：土木工程相关专业与学科领域

土木工程是一门诞生于石器时代的古老工程学科。美国土木工程师学会将土木工程定义为一门致力于将习得的数学、物理科学知识、实践经验以及判断能力应用于发展更经济的使用材料以及自然资源的方法，以改善人类福祉的专业。

本章将论述土木工程所囊括的不同部分，并介绍国家基础设施建设的重要意义。

土木工程涉及下述专业与学科领域：

(i) 测量学

（ii）建筑材料
（iii）建筑工艺学
（iv）结构工程
（v）岩土工程
（vi）水力学
（vii）水资源及灌溉工程
（viii）运输工程
（ix）环境工程
（x）城市规划

各领域包含内容如下：

（i）测量学

测量学是有关地图绘制的科学。在进行任何一项工程活动前都需要确定各物体的水平和竖向相对位置。这将通过场地测量实现。过去传统的测量设备包括皮尺、水准仪等。而在现今的电子时代，测距仪、全站仪等现代测量设备可以更方便、更准确地获取测量结果。摄影测量学及遥感等现代技术也将使测量工作变得更为容易。

（ii）建筑材料

遮风避雨是文明的基本需求。自文明诞生起就不断致力于建造更好的庇护场所。传统建筑材料包括石材、砖块、木材、石灰、水泥、沙子、凝胶及瓷砖等。采用钢、铝、玻璃、釉面砖、熟石膏、油漆、清漆提升了建筑质量。采用合适的配比可以将石灰、水泥及沙子配置为砂浆。水泥、沙子、凝胶（碎石）和水可以混合产生混凝土。在混凝土合适位置配置钢筋有助于建造结实耐久的高层建筑。钢筋和混凝土的混合材料被称为钢筋水泥混凝土，工程通称为RCC。土木工程师需要了解所有的建筑材料性质才能适当地使用各种材料。市场不断出现新升级的建筑材料，优秀的土木工程师会率先使用新型建材。

（iii）施工技术

施工是土木工程的主要活动，正在不断获得发展。由于市区场地成本不断攀升，对高层建筑的需求正在不断增长，而郊区需要较低的建设成本。土木工程师应当根据可用建筑材料采用合适的建造技术。

（iv）结构工程

结构所承担的荷载最终会传导到地面。这个过程中，结构构件将承受内力。例如，建筑结构中楼板所承担的荷载将通过梁、柱、基础传导到地面。确定结构构件内力的过程称为结构分析，选取结构构件合适尺寸的过程称为结构设计。需要分析的结构包括砌体结构、钢筋混凝土结构和钢结构。20世纪60年代中期分析方法获得了显著进步。伴随着计算机出现，数值分析方法也产生了，同时设计程序也得到了普及。矩阵分析方法和有限元分析方法为复杂结构的分析提供了辅助。土木工程师不仅要考虑结构的安全性，还要考虑经济性。为了得到经济的设计方案，数学优化方法得到了应用。近年来地震频发，因此对地震力的分析也十分重要。设计抗震结构引起了大批研究者的关注。这些都属于结构工程的范畴。

（v）岩土工程

各地的土的性质各不相同，甚至位置相同但不同深度的土的性质也不相同。随着季节变换，土中水分含量不同也会导致土的性质发生变化。结构荷载最终要安全地传导到土上。因此，需要适当地评定土的安全承载能力。土木工程中的这一专业称为岩土工程。

除确定建筑地基安全承载能力外，岩土工程还涉及路面、隧道、堤坝、运河及挡土结构的设计。岩土工程还包括地基处理技术。

(vi) 水力学

万物生长需要水。水的力学性质以及水流的性质是土木工程研究的另一个重要领域，这被称为水力学。

(vii) 水资源和灌溉工程

水会供应于农业灌溉及居民饮用。因此需要识别和存储水资源。水资源和灌溉工程包括对水资源的识别、规划，包括建设水池、水坝等蓄水结构，以及将蓄积的水源运输至使用地。

(viii) 运输工程

交通运输设施是另一个重要需求。提供良好而经济的道路是土木工程师的职责。这包括设计路基、路面、交叉排水工程、交叉路口、涵洞、桥梁、隧道等等。铁路运输是另一种重要的长途运输设施。设计、建造和维护铁路线路及信号系统也属于交通运输工程。同样，有对机场和港口的需求。为建设此类交通运输设施，需要进行交通调查。交通调查和道路、桥梁、铁路、码头机场的设计、建设及维护称为运输工程。

(ix) 环境工程

将水合理地分配给乡村和城市、处理废水和固体垃圾，属于土木工程的另一领域。工业化及车辆交通会产生空气污染问题。环境工程通过处理这些问题来给公众提供健康的环境。

(x) 建筑及城市规划

建筑需要符合美学，城乡需要合理规划。这些工程领域已经得到了极大的发展，业已从土木工程领域分离出来，成为单独的学科。

Text B Infrastructure

Infrastructure facilities involve various civil engineering **amenities**: electricity, telephone, internet facility, educational and healthcare facilities. Civil engineering amenities in the infrastructure developments are listed below:

(i) A good town planning and developing sites

(ii) Providing suitable roads and network of roads

(iii) Railway connection to important places

(iv) Airports of national and international standards

(v) Assured water supply to towns, cities and rural areas

(vi) A good drainage and waste disposal system

(vii) Pollution free environment

Connecting producing centre to marketing places minimises exploitation from middlemen. Both producer and consumers are benefitted. Imports and exports become easy as a result of which whole world becomes a village. The infrastructure development generates scope for lots of industries. Manpower is utilized for the benefit of mankind. Antisocial activities come under control. Improved education and healthcare give rise to skilled and healthy work force. Quality of life of the people is improved. In case of **natural calamities** assistance can be extended easily and misery of affected people is reduced. Infrastructure facility improves defence system and peace exists in the country. Improved economical power of the country brings a respectable status in the world.

The world has realized that a government should not involve itself in production and distribution but should develop infrastructure to create an atmosphere for economical development.

A civil engineer has to conceive, plan, estimate, get approval, create and maintain all civil engineering infrastructure activities. He has to carry out research and training programmes to improve the technology. Civil engineer has a very important role in the development of the following infrastructures:

(i) Town and city planning.

(ii) Build suitable structures for the rural and urban areas for various utilities.

(iii) Build **tanks, dams** to exploit water resources.

(iv) Purify the water and supply water to needy areas like houses, schools, offices, and agriculture field.

(v) Provide good **drainage system** and **purification plants.**

(vi) Provide and maintain communication systems like roads, railways, harbours and airports.

(vii) Monitor land, water and air pollution and take measures to control them.

New Words and Expressions

[1] amenity [ə'miːnəti] *n.* 设施；

[2] natural calamities 自然灾害；

[3] tanks [tæks] *n.* 罐，槽；

[4] dam [dæm] *n.* 水坝；

[5] drainage system 排水系统；

[6] purification plants 净化厂。

LESSON 2   SOIL MECHANICS

Text A   The Origin of Soils

To the civil engineer, soil is any **uncemented** or weakly cemented accumulation of mineral particles formed by the weathering of rocks as part of the rock cycle (Fig. 2.1), the void space between the particles containing water and/or air. Weak cementation can be due to **carbonates** or oxides **precipitated** between the particles, or due to organic matter. Subsequent deposition and compression of soils, combined with cementation between particles, transforms soils into **sedimentary** rocks (a process known as **lithification**). If the products of weathering remain at their original location, they constitute a residual soil. If the products are transported and deposited in a different location, they constitute a transported soil, the agents of transportation being gravity, wind, water and glaciers. During transportation, the size and shape of particles can undergo change and the particles can be sorted into specific size ranges.

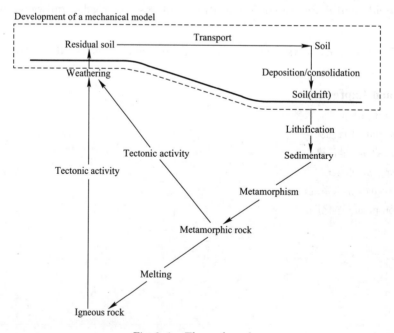

Fig. 2.1   The rock cycle

Particle sizes in soils can vary from over 100mm to less than 0.001mm. In the UK, the size ranges are described as shown in Fig. 2.2. In Fig. 2.2, the terms 'clay', 'silt' etc. are used to describe only the sizes of particles between specified limits. However, the same terms are also used to describe particular types of soil, classified according to their mechanical behaviours.

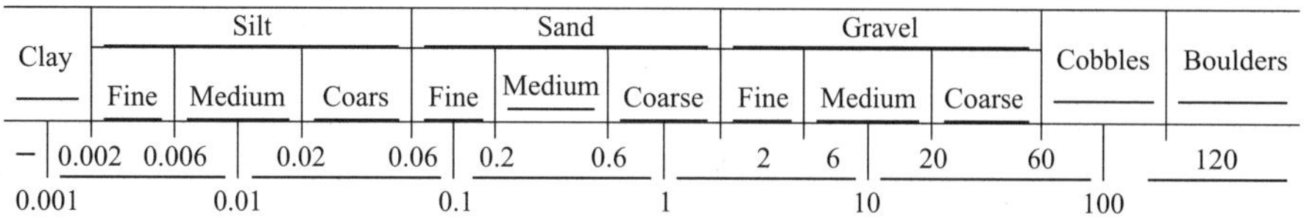

Clay	Silt			Sand			Gravel			Cobbles	Boulders
	Fine	Medium	Coars	Fine	Medium	Coarse	Fine	Medium	Coarse		
—	0.002	0.006	0.02	0.06	0.2	0.6	2	6	20	60	120
0.001		0.01			0.1		1		10	100	

Fig. 2.2 Particle size ranges

The type of transportation and subsequent deposition of soil particles has a strong influence on the distribution of particle sizes at a particular location. Some common depositional regimes are shown in Fig. 2.3. In glacial regimes, soil material is eroded from underlying rock by the frictional and freeze-thaw action of glaciers. The material, which is typically very varied in particle size from clay to boulder-sized particles, is carried along at the base of the glacier and deposited as the ice melts, the resulting material is known as (**glacial**) **till**. Similar material is also deposited as a terminal moraine at the edge of the glacier. As the glacier melts, moraine is transported in the outwash; it is easier for smaller, lighter particles to be carried in suspension, leading to a **gradation** in particle size with distance from the glacier as shown in Fig. 2.3 (*a*). In warmer temperate climates the chief transporting action is water (i.e. rivers and seas), as shown in Fig. 2.3 (*b*). The deposited material is known as **alluvium**, the composition of which depends on the speed of water flow. Faster-flowing rivers can carry larger particles in suspension, resulting in alluvium, which is a mixture of sand and gravel-sized particles, while slower-flowing water will tend to carry only smaller particles. At estuarine locations where rivers meet the sea, material may be deposited as a shelf or **delta**. In arid (desert) environments (Fig. 2.3*c*) wind is the key agent of transportation, eroding rock outcrops and forming a **pediment** (the desert floor) of fine wind-blown sediment (**loess**). Towards the coast, a **playa** of temporary evaporating lakes, leaving salt deposits, may also be formed. The large temperature differences between night and day additionally cause thermal weathering of rock outcrops, producing **scree**. These surface processes are geologically very recent, and are referred to as **drift deposits** on geological maps.

The relative proportions of different-sized particles within a soil are described as its particle size distribution (PSD), and typical curves for materials in different depositional environments are shown in Fig. 2.4.

At a given location, the subsurface materials will be a mixture of rocks and soils, stretching back many hundreds of millions of years in geological time. As a result, it is important to understand the full geological history of an area to understand the likely characteristics of the deposits that will be present at the surface, as the depositional regime may have changed significantly over geological time. As an example, the West Midlands in the

Basic characteristics of soils

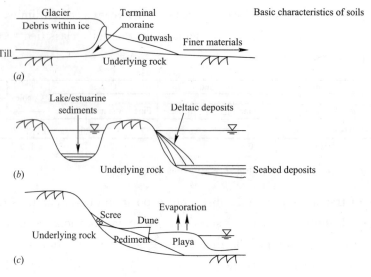

Fig. 2.3 Common depositional environments
(a) Glacial; (b) Fluvial; (c) Desert

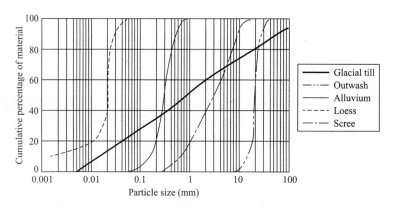

Fig. 2.4 Particle size distributions of sediments from
different depositional environments

UK was deltaic in the Carboniferous period (395~345 million years ago), depositing organic material which subsequently became coal measures. In the subsequent Triassic period (280~225 million years ago), due to a change in sea level sandy materials were deposited which were subsequently **lithified** to become Bunter sandstone. Mountain building during this period on what is now the European continent caused the existing rock layers to become folded. It was subsequently flooded by the North Sea during the Cretaceous/Jurassic periods (225~136 million years ago), depositing fine particles and carbonate material (Lias clay and Oolitic limestone). The Ice Ages in the Pleistocene period (1.5~2 million years ago) subsequently led to glaciation over all but the southernmost part of the UK, eroding some of the recently deposited softer rocks and depositing glacial till. The subsequent melting of the glaciers created river valleys, which deposited alluvium above the till. The geological history would therefore suggest that the surficial soil conditions are likely to consist of alluvium overlying till/clay overlying stronger rocks, as shown schematically in Fig. 2.5. This example demonstrates the importance of engineering geology in understanding ground conditions. A thorough introduction to this topic can be found in Waltham (2002).

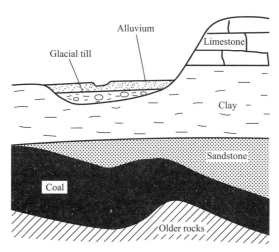

Fig. 2.5　Typical ground profile in the West Midlands, UK

The destructive process in the formation of soil from rock may be either physical or chemical. The physical process may be erosion by the action of wind, water or glaciers, or disintegration caused by cycles of freezing and thawingin cracks in the rock. The resultant soil particles retain the same mineralogical composition as that of the parent rock (a full description of this is beyond the scope of this book). Particles of this type are described as being of 'bulky' form, and their shape can be indicated by terms such as angular, rounded, flat and elongated. The particles occur in a wide range of sizes, from boulders, through gravels and sands, to the fine rock flour formed by the **grinding** action of glaciers. The structural arrangement of **bulky** particles is described as **single grain**, each particle being in direct.

New Words and Expressions
[1]　uncemented ['ʌnsɪ'mentɪd] *adj.* 未胶结的；
[2]　carbonate ['kɑːbəneɪt] *n.* 碳酸盐；
[3]　precipitate [prɪ'sɪpɪteɪt] *vt.* 使沉淀，使凝结；*vi.* 沉淀，凝结；*n.* 沉淀物；
[4]　sedimentary [ˌsedɪ'mentrɪ] *adj.* 沉积的；
[5]　lithification [ˌlɪθɪfɪ'keɪʃən] *n.* 岩化；
[6]　glacial ['gleɪʃl] *adj.* 冰的，冰河［川］的；*n.* 冰川；
[7]　till [tɪl] *n.* 冰碛土；
[8]　alluvium [ə'luːvɪəm] *n.* 冲积层，淤积层；
[9]　gradation [grə'deɪʃn] *n.* 级配；

[10] delta ['deltə] n. （河流的）三角洲；
[11] pediment ['pedɪmənt] n. 山形墙，三角墙；山墙；人形山头；
[12] loess ['ləʊɪs] n. 黄土；
[13] playa ['plaɪə] n. 沙漠中的盆地，海边休养地；
[14] scree [skriː] n. 卵石；
[15] drift deposit 堆积沉淀物；
[16] lithify ['lɪθɪfaɪ] v. （使）岩化；
[17] grind [graɪnd] vt. & vi. 磨碎；
[18] bulky ['bʌlkɪ] adj. 庞大的，笨重的，体积大的；
[19] single grain 单颗粒，单粒。

参考译文：土体的来源

对土木工程师来说，土是指岩石在风化过程中形成的无粘结或弱粘结沉积物（图2.1）。土体固相颗粒之间的孔隙由水或气填充。碳酸盐、氧化物或有机质的沉淀凝结导致了土体固相颗粒之间的弱胶结效应。土体的沉积、压缩以及颗粒间的胶结作用可以将土转化为沉积岩（此过程称为岩化）。岩石风化产物如果残留于原位，则称为残积土。风化产物若被转移至新位置沉积，则称为运积土。搬运风化产物的方式有重力、风、水和冰川。在搬运过程中，土颗粒的大小、形状可能发生改变，可以将土颗粒按照不同粒径尺寸进行划分。

颗粒粒径尺寸范围分布广泛，有的大于100mm，有的小于0.001mm。图2.2为英国典型土颗粒粒径分布范围。图2.2中"黏土"、"淤泥"等用于指代界限粒径范围内的土颗粒。然而，这些术语也用于指代特定类型的土壤，这些类型是根据土的力学性质划分的。

土颗粒的搬运及沉积方式对某一场地的土的粒径级配有显著影响。图2.3显示了一些常见的沉积方式。在冰川沉积过程中，土体材料是被冰川通过摩擦和冻融作用从下卧的岩层中侵蚀出来的。土颗粒粒径分布范围较大，小至黏粒，大至漂石，这些颗粒随着冰川移动，随着冰川融化而沉积，这种方式形成的土体称为冰碛土。类似的土体材料也会在冰川边缘沉积为冰碛。随着冰川融化，冰碛转化为冰水停积；小的轻的颗粒更容易悬浮在上层随水迁移，因此产生图2.3(a)所示的随距冰川距离变化的粒径级配分布。如图2.3(b)所示，在气候温暖的地方主要的搬运作用来自于水（例如河流和海洋）。形成的土体称为冲积土，冲积土的组分与水流速度有关。流速大的河流可以裹挟更大尺寸的土颗粒，形成混合了砂与砾石的冲积土。而流速慢的水流仅能携带较细的土颗粒。在河流入海口，土体材料将沉积成大陆架或三角洲。在干旱（沙漠）环境中（图2.3c），风是土体搬运的主要媒介，风侵蚀露出的岩石，产生的细小的黄土构成沙漠的地表。在海滨或临时蒸干湖边会留下盐的沉积物。昼夜巨大温差变化会导致岩石的温度风化，形成卵石。这些地表风化过程发生在地质史的近代，在地质图上标注为漂移沉积物。

土体中不同大小颗粒相对比例关系可用颗粒级配表示，图2.4显示了不同沉积环境的典型土的颗粒级配曲线。

在一个给定的场地，地表以下由岩石、土体混合而成的，从地质年代上看已经经历了数亿年。因此，了解这个区域的整个地质历史对于掌握地表沉积物的特性非常重要，因为沉积方式在地质年代中可能产生了显著变化。例如，英国西米德兰兹郡在石炭纪（距今

3.95~3.45亿年）为冲积三角洲，沉积的有机物后续形成了煤炭。在之后的三叠纪（距今2.8~2.25亿年）海平面变化导致沉积物主要为砂类物质，后续岩化为杂色砂岩。这一时期在如今的欧洲大陆的造山运动引起了岩层的褶皱。之后，在白垩纪和侏罗纪（距今2.25~1.36亿年）该地被北海淹没，沉积物为细小的碳酸盐（青色石灰岩和鲕状灰岩）。在更新世的冰河时期（距今150~200万年），冰川覆盖了除英国最南部外大部分区域，侵蚀掉了部分新形成的软岩并沉积为冰碛土。之后，冰川融化形成河谷，在冰碛土形成冲积层。因此，地质历史表明基岩之上的地表土层很可能包括冲积土、冰碛土\黏土，如图2.5所示。这个例子表明了工程地质对于掌握场地条件的重要意义。对此问题更详尽的介绍参考Waltham（2002）。

将岩石破坏形成土体的风化过程包括物理风化和化学风化。物理风化可能是由风、水或冰川，或岩石裂缝中的水的冻融循环产生的侵蚀作用。产生的土颗粒的矿物成分和母岩的成分相同（对此的完整解释超出了本书范围）。由此形成的土颗粒粒径较大，它们的形状可能是颗粒状、圆形、扁平和长条形。土颗粒粒径分布较广，从漂石到砾石，到砂，再到冰川磨碎产生的石粉。大块土颗粒的结构分布用直接接触的单个颗粒表示。

Text B Vertical Stresses in a Layer

In many places on earth the soil consists of practically horizontal layers. If such a soil does not carry a local surface load, and if the groundwater is at rest, the vertical stresses can be determined directly from a consideration of vertical equilibrium. The procedure is illustrated in this chapter.

A simple case is a homogeneous layer, completely saturated with water, see Fig. 2.6. The pressure in the water is determined by the location of the **phreatic** surface. This is defined as the plane where the pressure in the groundwater is equal to the atmospheric pressure. If the atmospheric pressure is taken as the zero level of pressures, as is usual, it follows that $p=0$ at the phreatic surface. If there are no **capillary** effects in the soil, this is also the upper boundary of the water, which is denoted as the groundwater table. In the example it is assumed that the phreatic surface coincides with the soil surface, see Fig. 2.6. The volumetric weight of the saturated soil is supposed to be $=20kN/m^3$. The vertical normal stress in the soil now increases linearly with depth,

$$\sigma_{zz} = \gamma d \tag{2.1}$$

This is a consequence of vertical equilibrium of a column of soil of height d. It has been assumed that there are no shear stresses on the vertical planes bounding the column in horizontal direction. That seems to be a reasonable assumption if the **terrain** is homogeneous and very large, with a single geological history. Often this is assumed, even when there are no data.

At a depth of 10m, for instance, the vertical total stress is $200kN/m^2 = 200kPa$. Because the groundwater is at rest, the pressures in the water will be **hydrostatic.** The soil can be considered to be a container of water of very complex shape, bounded by all the particles, but that is irrelevant for the actual pressure in the water. This means that the pressure in the water at a depth d will be equal to the weight of the water in a column of unit area, see also Fig. 2.6,

$$p = \gamma_w d \qquad (2.2)$$

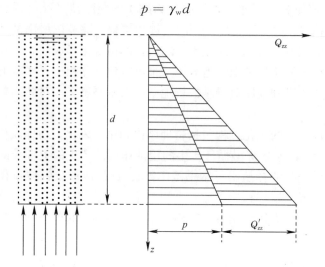

Fig. 2.6 Stresses in a homogeneous layer

where γ_w is the volumetric weight of water, usually $\gamma_w = 10 \text{kN/m}^3$. It now follows that a depth of 10m the effective stress is 200kPa−100kPa=100kPa.

Formally, the distribution of the effective stress can be found from the basic equation $\sigma'_{zz} = \sigma_{zz} - p$, or, with (2.1) and (2.2),

$$\sigma'_{zz} = (\gamma - \gamma_w) d \qquad (2.3)$$

The vertical effective stresses appear to be linear with depth. That is a consequence of the linear distribution of the total stresses and the pore pressures, with both of them being zero at the same level, the soil surface.

It should be noted that the vertical stress components, both the total stress and the effective stress, can be found using the condition of vertical equilibrium only, together with the assumption that the shear stresses are zero on vertical planes. The horizontal normal stresses remain undetermined at this stage. Even by also considering horizontal equilibrium these horizontal stresses can not be determined. A consideration of horizontal equilibrium, see Fig. 2.7, does give some additional information, namely that the horizontal normal stresses on the two vertical planes at the left and at the right must be equal, but their **magnitude** remains unknown. The determination of horizontal (or lateral) stresses is one of the essential difficulties of soil mechanics. Because the horizontal stresses can not be determined from equilibrium conditions they often remain unknown. It will be shown later that even when also considering the deformations, the determination of the horizontal stresses remains very difficult, as this requires detailed knowledge of the geological history, which is usually not available. Perhaps the best way to determine the horizontal stresses is by direct or indirect measurement in the field. The problem will be discussed further in later chapters.

The simple example of Fig. 2.6 may be used as the starting point for more complex cases. As a second example the situation of a somewhat lower phreatic surface is considered, say when it is lowered by 2m. This may be caused by the action of a pumping station in the area, such that the water level in the **canals** and the **ditches** in a **polder** is to be kept at a level of 2m below the soil surface. In this case there are two possibilities, depending upon the size of the particles in the soil. If the soil consists of very coarse material, the groundwater

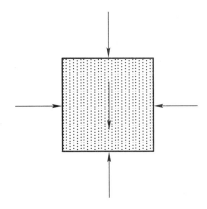

Fig. 2.7 Equilibrium

level in the soil will coincide with the phreatic surface (the level where $p=0$), which will be equal to the water level in the open water, the ditches. However, when the soil is very fine (for instance clay), it is possible that the top of the groundwater in the soil (the groundwater level) is considerably higher than the phreatic level, because of the effect of *capillarity*. In the fine pores of the soil the water may rise to a level above the phreatic level due to the suction caused by the surface tension at the interface of particles, water and air. This surface tension may lead to pressures in the water below atmospheric pressure, i.e. negative water pressures. The zone above the phreatic level is denoted as the *capillary zone*. The maximum height of the groundwater above the phreatic level is denoted as h_c, the *capillary rise*.

If the capillary rise h_c in the example is larger than 2m, the soil in the polder will remain saturated when the water table is lowered by 2m. The total stresses will not change, because the weight of the soil remains the same, but the pore pressures throughout the soil are reduced by $\gamma_w \times 2m = 20 kN/m^2$. This means that the effective stresses are increased everywhere by the same amount, see Fig. 2.8.

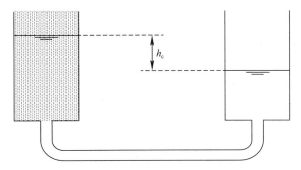

Fig. 2.8 Capillary rise

Lowering the phreatic level appears to lead to an increase of the effective stresses. In practice this will cause deformations, which will be manifest by a subsidence of the ground level. This indeed occurs very often, wherever the groundwater table is lowered. Lowering the water table to construct a dry building pit, or lowering the ground-water table in a newly reclaimed polder, leads to higher effective stresses, and therefore settlements. This may be accompanied by severe damage to buildings and houses, especially if the settlements are not uniform. If the subsidence is uniform there is less risk for damage to structures founded on the soil in that area.

Lowering the phreatic level may also have some positive consequences. For instance, the increase of the effective stresses at the soil surface makes the soil much **stiffer** and stronger, so that heavier vehicles (tractors or other agricultural machines) can be supported. In case of a very high phreatic surface, coinciding with the soil surface, as illustrated in Figure 2.9, the effective stresses at the surface are zero, which means that there is no force between the soil particles. Man, animal and machine then can not find support on the soil, and they may sink into it. The soil is called **soggy** or **swampy**. It seems natural that in such cases people will be motivated to lower the water table. This will result in some subsidence, and thus part of the effect of the lower groundwater table is lost. This can be restored by a further lowering of the water table, which in turn will lead to further subsidence. In some places on earth the process has had almost **catastrophic** consequences (Venice, Bangkok). The subsidence of Venice, for instance, was found to be caused for a large part by the production of ever increasing amounts of drinking water from the soil in the immediate vicinity of the city. Further subsidence has been reduced by finding a water supply farther from the city.

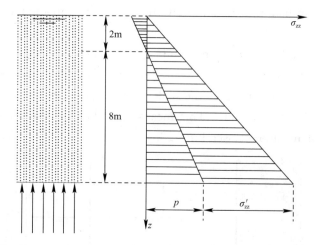

Fig. 2.9 Lowering the phreatic surface by 2m, with capillary rise

When the soil consists of very coarse material, there will practically be no capillarity. In that case lowering the phreatic level by 2m will cause the top 2m of the soil to become dry, see Fig. 2.10. The upper 2m of soil then will become lighter. A reasonable value for the dry volumetric weight is $\gamma_d = 16 \text{kN/m}^3$. At a depth of 2m the vertical effective stress now is $\sigma'_{zz} = 32\text{kPa}$, and at a depth of 10m the effective stress is $\sigma'_{zz} = 112\text{kPa}$. It appears that in this case the effective stresses increase by 12kPa, compared to the case of a water table coinciding with the ground surface. The distribution of total stresses, effective stresses and pore pressures is shown in Fig. 2.10. Again there will be a tendency for settlement of the soil. In later chapters a procedure for the calculation of these settlements will be presented. For this purpose first the relation between effective stress and deformation must be considered.

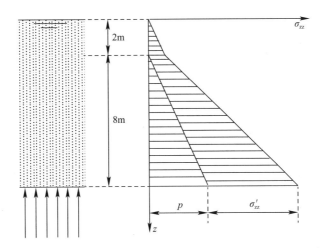

Fig. 2.10 Lowering of the phreatic surface by 2m, no capillarity

Subsidence of the soil can also be caused by the extraction of gas or oil from soil layers. The reservoirs containing oil and gas are often located at substantial depth (in Groningen at 2000m depth). These reservoirs usually consist of porous rock, that have been consolidated through the ages by the weight of the soil layers above it, but some porosity (say 10% or 20%) remains, filled with gas or oil. When the gas or oil is extracted from the reservoir, by reducing the pressure in the fluid, the effective stresses increase, and the thickness of the reservoir will be reduced. This will cause the soil layers above the reservoir to settle, and it will eventually give rise to subsidence of the soil surface. In Groningen the subsidence above the large gas reservoir is estimated to reach about 50cm, over a very large area. All structures subside with the soil, with not very much risk of damage, as there are no large local variations to be expected. However, because the soil surface is below sea level, great care must be taken to maintain the drainage capacity of the hydraulic infrastructure. Sluices may have to be renewed because they subside, whereas water levels must be maintained. The dikes also have to be raised to balance the subsidence due to gas production.

In some parts of the world subsidence may have very serious consequences, for instance in areas of coal mining activities. In mining the entire soil is being removed, and sudden collapse of a mine gallery may cause great damage to the structures above it.

It has been indicated in the examples given above how the total stresses, the effective stresses and the pore pressures can be determined on a horizontal plane in a soil consisting of practically horizontal layers. In most cases the best general procedure is that first the total stresses are determined, from the vertical equilibrium of a column of soil. The total stress then is determined by the total weight of the column (particles and water), plus an eventual **surcharge** caused by a structure. In the next step the pore pressures are determined, from the hydraulic conditions. If the groundwater is at rest it is sufficient to determine the location of the phreatic surface. The pore pressures then are hydrostatic, starting from zero at the level of the phreatic surface, i. e. linear with the depth below the phreatic surface. When the soil is very fine a capillary zone.

New Words and Expressions

[1] phreatic [frɪ'ætɪk] *adj*. 地下水的；
[2] capillary ['kæpəleri;] *n*. 毛细管；*adj*. 毛状的；表面张力的；
[3] terrain [tə'reɪn] *n*. 地形，地势；地面，地带；
[4] hydrostatic [ˌhaɪdrə'stætɪk] *adj*. 静水力学的，流体静力学的；
[5] magnitude ['mægnɪtjuːd] *n*. 巨大；重要；量级；（地震）级数；
[6] canal [kə'næl] *n*. 运河，沟渠；管道；
[7] ditch [dɪtʃ] *n*. 沟渠；壕沟；
[8] polder ['pəʊldə] *n*. 开拓地（尤指荷兰等国围海造的低田）；圩垸；
[9] stiff [s'tɪfər] *adj*. 僵硬的；
[10] soggy ['sɒɡɪ] *adj*. 湿透的，浸透的，潮湿的；
[11] swampy ['swɒmpɪ] *adj*. 沼泽的；湿软的；
[12] catastrophic [ˌkætə'strɒfɪk] *adj*. 灾难的；惨重的；
[13] surcharge ['sɜːtʃɑːdʒ] *vt*. 使……超载；*n*. 附加费；超载。

LESSON 3 ENGINEERING MECHANICS

Text A Types of Beams, Loads, and Reactions

Structural members are usually classified according to the types of loads that they support. For instance, an **axially** loaded bar supports forces having their vectors directed along the axis of the bar, and a bar in **torsion** supports **torques** (or couples) having their moment vectors directed along the axis. In this chapter, we begin our study of beams (Fig. 3.1), which are structural members subjected to lateral loads, that is, forces or moments having their vectors perpendicular to the axis of the bar.

The beams shown in Fig. 3.1 are classified as **planar** structures because they lie in a single plane. If all loads act in that same plane, and if all deflections occur in that plane, then we refer to that plane as the plane of bending.

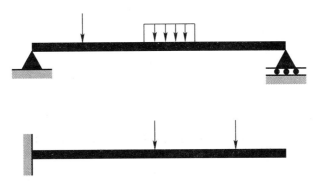

Fig. 3.1 Examples of beams subjected to lateral loads

In this chapter we discuss shear forces and bending moments in beams, and we will show how these quantities are related to each other and to the loads. Finding the shear forces and bending moments is an essential step in the design of any beam. We usually need to know not only the maximum values of these quantities, but also the manner in which they vary along the axis. Once the shear forces and bending moments are known, we can find the stresses, strains, and deflections, as discussed later in Chapters 5, 6, and 9.

Beams are usually described by the manner in which they are supported. For instance, a beam with a pin support at one end and a roller support at the other (Fig. 3.2a) is called a **simply supported beam** or a simple beam. The essential feature of a **pin support** is that it prevents translation at the end of a beam but does not prevent **rotation.** Thus, end A of the beam of Fig. 3.2 (a) cannot move horizontally or vertically but the axis of the beam can rotate in the plane of the figure. Consequently, a pin support is capable of developing a force reaction with both horizontal and vertical components (H_A and R_A), but it cannot develop a moment reaction.

At end B of the beam (Fig. 3.2a) the **roller support** prevents translation in the vertical direction but not in the horizontal direction; hence this support can resist a vertical force (R_B) but not a horizontal force. Of course, the axis of the beam is free to rotate at B just as it is at A. The vertical reactions at roller supports and pin supports may act *either* upward or downward, and the horizontal reaction at a pin support may act either to the left or to the right. In the figures, reactions are indicated by slashes across the arrows in order to distinguish them from loads.

The beam shown in Fig. 3.2 (*b*), which is fixed at one end and free at the other, is called a **cantilever beam**. At the **fixed support** (or *clamped support*) the beam can neither translate nor rotate, whereas at the free end it may do both. Consequently, both force and moment reactions may exist at the fixed support.

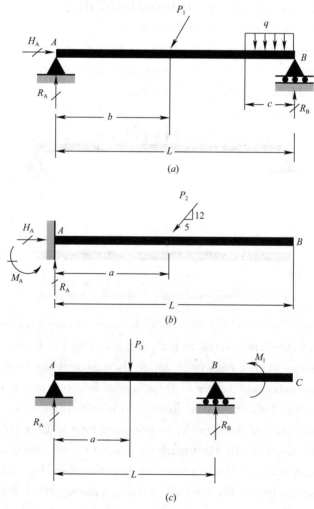

Fig. 3.2 Types of beams

(*a*) Simple beam; (*b*) Cantilever beam; (*c*) Beam with an overhang

The third example in the figure is a **beam with an overhang** (Fig. 3. 2c). This beam is simply supported at points A and B (that is, it has a pin support at A and a roller support at B) but it also projects beyond the support at B. The overhanging segment BC is similar to a cantilever beam except that the beam axis may rotate at point B.

When drawing sketches of beams, we identify the supports by conventional symbols, such as those shown in Fig. 3. 2. These symbols indicate the manner in which the beam is restrained, and therefore they also indicate the nature of the reactive forces and moments. However, *the symbols do not represent the actual physical construction*. For instance, consider the examples shown in Fig. 3. 3. Part (a) of the figure shows a **wide-flange** beam supported on a concrete wall and held down by anchor bolts that pass through **slotted** holes in the lower flange of the beam. This connection restrains the beam against vertical movement (either upward or downward) but does not prevent horizontal movement. Also, any restraint against rotation of the longitudinal axis of the beam that is small and ordinarily may be disregarded. Consequently, this type of support is usually represented by a roller, as shown in part (b) of the figure.

The second example (Fig. 3. 3c) is a beam-to-column connection in which the beam is attached to the column flange by bolted angles. This type of support is usually assumed to restrain the beam against horizontal and vertical movement but not against rotation (restraint against rotation is slight because both the angles and the column can bend). Thus, this connection is usually represented as a pin support for the beam (Fig. 3. 3d).

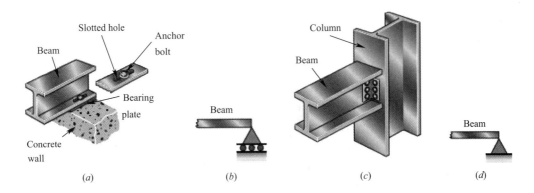

Fig. 3. 3 Beam supported on a wall (1)
(a) Actual construction; (b) Representation as a roller support. Beam-to-column connection;
(c) Actual construction, and (d) Representation as a pin supp0ort

The last example (Fig. 3. 3e) is a metal pole **welded** to a base plate that is anchored to a concrete pier **embedded** deep in the ground. Since the base of the pole is fully restrained against both translation and rotation, it is represented as a fixed support (Fig. 3. 3f).

The task of representing a real structure by an **idealized model**, as illustrated by the beams shown in Fig. 3. 2, is an important aspect of engineering work. The model should be

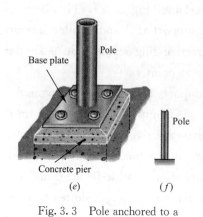

Fig. 3.3 Pole anchored to a concrete pier (2)
(e) Actual construction;
(f) Representationas a fixed support

simple enough to facilitate mathematical analysis and yet complex enough to represent the actual behavior of the structure with reasonable accuracy. Of course, every model is an approximation to nature. For instance, the actual supports of a beam are never perfectly rigid, and so there will always be a small amount of translation at a pin support and a small amount of rotation at a fixed support. Also, supports are never entirely free of **friction**, and so there will always be a small amount of restraint against translation at a roller support. In most circumstances, especially for statically determinate beams, these deviations from the idealized conditions have little effect on the action of the beam and can safely be disregarded.

New Words and Expressions

[1] axially ['æksɪəlɪ] adv. 轴向；
[2] torsion ['tɔːʃn] n. 扭转；扭曲；
[3] torques ['tɔːkwiːz] n. （尤指机器的）扭转力（torque 的名词复数）；转（力）矩；
[4] planar ['pleɪnə(r)] adj. 平面的，平坦的；
[5] simply supported beam 简支梁；
[6] pin support [pin səˈpɔːt] 铰支座；
[7] rotation [rəʊˈteɪʃn] n. 旋转，转动；轮流，循环；
[8] roller support ['rəʊlə səˈpɔːt] 滚动支座；
[9] cantilever beam ['kæntiliːvəbiːm] 悬（臂）梁；悬臂梁；
[10] fixed support [fɪkst səˈpɔːt] 固定支座；
[11] beam with an overhang 伸臂梁；
[12] wide-flange ['waɪdflˈændʒ] 宽缘的；宽凸缘；
[13] slotted ['slɒtɪd] adj. 开有槽沟的；v. 把……放入狭长开口中；
[14] welded ['weldɪd] v. 焊接（weld 的过去式和过去分词）；熔接；锻接；
[15] embedded [ɪmˈbedɪd] adj. 植入的，深入的，内含的；v. 把……嵌入，埋入；
[16] idealized model 理想化模型，理想模式；
[17] friction ['frɪkʃn] n. 摩擦；摩擦力。

参考译文：梁、荷载及反力类别

结构构件通常根据其承受的荷载进行分类。例如，一根轴向受力杆件所承受的力的方向为沿着轴向，而一根受扭杆件所承受的扭矩（或力偶）为垂直于轴向。本章从梁式构件入手（图3.1），梁是一种承受横向荷载的结构构件，也就是说承受的力和力矩垂直于杆件轴向。

图 3.1 中所示的梁存在于某一平面，属于平面结构。如果所有的荷载和变形都发生在该平面内，我们称该平面为弯曲平面。

本章将论述梁中的剪力和弯矩，我们会展示剪力与弯矩之间的关联及其与外荷载的关系。确定剪力、弯矩的大小是梁设计中必不可少的一步。我们通常不仅需要知道这些量的最大值，还需要知道它们沿轴向的变化规律。一旦剪力和弯矩确定了，我们就可以按照后续第 5、6 和 9 章内容确定应力、应变及变形。

梁通常按照支撑方式来命名。例如，一端铰支另一端滚动支座的梁称为简支梁（图 3.2a）。铰支座本质的特点是约束了梁一端的位移但并不约束转角。因此，图 3.2(a) 中梁 A 端不能发生水平或竖向移动，但梁的轴线可以在平面内发生转动。因此，铰支座可以产生水平和竖直方向的反力（H_A 和 R_A），但不能产生弯矩。

图 3.2a 中梁 B 端的滚动支座可以防止竖直方向的位移，但不能限制水平方向的位移，因此这种支座可以产生竖向反力 R_B，但不产生水平反力。当然梁的 B 端和 A 端同样可以自由旋转。铰支座和滚动支座的竖向反力可以是向上的，也可能是向下的，铰支座的水平反力可能向左，也可能向右。图中的反力标注上了斜线以免和荷载混淆。

图 3.2b 所示的梁一端固定，另一端自由，这样的梁被称为悬臂梁。在固定支座（或嵌固端）梁不能平移也不能转动，而在悬臂端既可以产生线位移又可以产生转动。因此在固定支座处会产生反力和弯矩。

图中第三种类型的梁有一个伸臂段（图 3.2c）。这种梁在 A 点和 B 点简支（即 A 点为铰支座，B 点为滚动支座），但梁有一部分伸出 B 点。外伸的 BC 段类似于悬臂梁，除了 B 点段可能有转角。

在绘制梁的图时我们采用图 3.2 中所示的常规符号来表示梁的支撑。这些符号表明了梁的约束以及支座反力支座弯矩的性质。然而，这些符号并不表示实际构造方式。例如图 3.3 中的几个示例，图 3.3(a) 中显示了一个宽翼缘的梁通过穿过下翼缘槽孔的螺栓锚固在混凝土墙上。这种连接形式约束梁的竖向位移（包括向上和向下），但并不限制水平向的运动。同样，这种约束对于转动的限制较小，可以忽略。因此这种约束通常表示滚动支座，如图 3.3(b) 所示。

图 3.3(c) 所示的第二个例子是一个梁柱节点，在该节点处梁通过螺栓和角钢连接在柱子翼缘上。通常假设这种连接可以抵抗水平方向和竖直方向的位移，但不能抵抗转动（由于角钢和柱子都可以弯曲，对转动的抵抗能力很小）。因此，这种连接形式一般标记成铰支座支撑。

最后一个例子（如图 3.3e 所示）是一个焊接在基础板上面的金属杆，基础板锚固在深埋入场地的混凝土墩上。由于杆的基础可以完全限制位移和转动，这种连接可以表示为固定支座。

图 3.2 展示了如何用理想模型来表示实际结构，这是工程中的一项重要工作。模型既需要足够简洁以便数学计算，又需要足够复杂，能以合理的精度反映结构的实际响应。当然，每个模型都是对本质的近似。例如，实际结构中梁的支撑从来都不是无限刚的，铰支座都会发生轻微位移，而固定支座都会有轻微的转动。同样，支座都存在一定摩擦，而滚动支座都会对位移有一定的限制。在大多数情况下，尤其是对静定梁而言，理想模型与实际情况的不同对梁的影响较小，可以忽略。

Text B Types of Loads and Reactions

Several types of loads that act on beams are illustrated in Fig. 3.2. When a load is applied over a very small area it may be idealized as a **concentrated load**, which is a single force. Examples are the loads P_1, P_2, and P_3 in the figure. When a load is spread along the axis of a beam, it is represented as a **distributed load**, such as the load q in part (a) of the figure. Distributed loads are measured by their **intensity**, which is expressed in units of force per unit distance (for example, newtons per meter or pounds per foot). A **uniformly distributed load**, or **uniform load**, has constant intensity q per unit distance (Fig. 3.2a). Avarying load has an intensity that changes with distance along the axis; for instance, the **linearly varying load** of Fig. 3.2 (b) has an intensity that varies linearly from q_1 to q_2. Another kind of load is a **couple**, illustrated by the couple of moment M_1 acting on the overhanging beam (Fig. 3.2c).

We assume in this discussion that the loads act in the plane of the figure, which means that all forces must have their vectors in the plane of the figure and all couples must have their moment vectors perpendicular to the plane of the figure. Furthermore, the beam itself must be symmetric about that plane, which means that every cross section of the beam must have a vertical axis of symmetry. Under these conditions, the beam will deflect only in the *plane of bending* (the plane of the figure).

Finding the reactions is usually the first step in the analysis of a beam. Once the reactions are known, the shear forces and bending moments can be found, as described later in this chapter. If a beam is supported in a statically determinate manner, all reactions can be found from free-body diagrams and equations of equilibrium.

In some instances, it may be necessary to add internal releases into the beam or frame model to better represent actual conditions of construction that may have an important effect on overall structure behavior. For example, the interior span of the bridge **girder** shown in Fig. 3.4 is supported on roller supports at either end, which in turn rest on reinforced concrete bents (or frames), but construction details have been inserted into the girder at either end to insure that the axial force and moment at these two locations are zero. This detail also allows the bridge deck to expand or contract under temperature changes to avoid inducing large thermal stresses into the structure. To represent these releases in the beam model, a **hinge** (or internal moment release, shown as a solid circle at each end) and an axial force release (shown as a C-shaped **bracket**) have been included in the beam model to show that both axial force (N) and bending moment (M), but not shear (V), are zero at these two points along the beam. (Representations of the possible types of releases for **two-dimensional** beam and torsion members are shown below the photo). As examples below shown, if axial, shear, or moment releases are present in the structure model, the structure should be broken into separate free-body diagrams (FBD) by cutting through the release; an additional equation of equilibrium is then available for use in solving for the unknown support reactions included in that FBD.

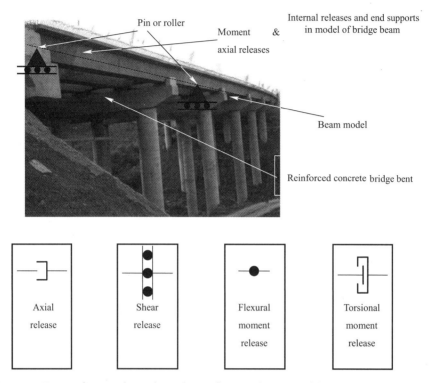

Fig. 3. 4 Types of internal member releases for two-dimensional beam and frame members

As an example, let us determine the reactions of the simple beam AB of Fig. 3. 5. This beam is loaded by an inclined force P_1, a vertical force P_2, and a uniformly distributed load of intensity q. We begin by noting that the beam has three unknown reactions: a horizontal force H_A at the pin support, a vertical force R_A at the pin support, and a vertical.

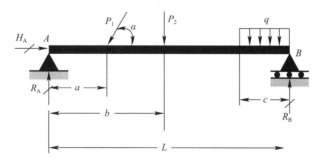

Fig. 3. 5 Simple beam (Repeated)

In finding this reaction we used the fact that the resultant of the distributed load is equal to the area of the **trapezoidal** loading diagram.

The moment reaction M_A at the fixed support is found from an equation of equilibrium of moments. In this example we will sum moments about point A in order to eliminate both H_A and R_A from the moment equation. Also, for the purpose of finding the moment of the distributed load, we will divide the trapezoid into two triangles, as shown by the dashed line in Fig. 3.2 (b). Each load triangle can be replaced by its resultant, which is a force having its magnitude equal to the area of the triangle and having its line of action through the **centroid** of the triangle. Thus, the moment about point A of the lower triangular part of the load is

$$\left(\frac{q_1 b}{2}\right)\left(L-\frac{2b}{3}\right) \tag{3.1}$$

in which $q_1 b/2$ is the resultant force (equal to the area of the triangular load diagram) and $L-2b/3$ is the moment arm (about point A) of the resultant.

The moment of the upper triangular portion of the load is obtained by a similar procedure, and the final equation of moment equilibrium (counterclockwise is positive) is

$$\Sigma M_A = 0 \quad M_A - \left(\frac{12 P_3}{13}\right) a - \frac{q_1 b}{2}\left(L - \frac{2b}{3}\right) - \frac{q_2 b}{2}\left(L - \frac{b}{3}\right) = 0 \tag{3.2}$$

from which

$$M_A = \frac{12 P_3 a}{13} + \frac{q_1 b}{2}\left(L - \frac{2b}{3}\right) + \frac{q_2 b}{2}\left(L - \frac{b}{3}\right) \tag{3.3}$$

Since this equation gives a positive result, the reactive moment M_A acts in the assumed direction, that is, counterclockwise. (The expressions for R_A and M_A can be checked by taking moments about end B of the beam and verifying that the resulting equation of equilibrium reduces to an identity.)

If the cantilever beam structure in Fig. 3.2 (b) is modified to add a roller support at B, it is now referred to as a one degree statically **indeterminate** "propped" cantilever beam. However, if a moment release is inserted into the model as shown in Fig. 3.6, just to the right of the point of application of load P_3, the beam can still be analyzed using the laws of statics alone because the release provides one additional equilibrium equation. The beam must be cut at the release to expose the internal stress resultants N, V, and M; now M=0 at the release so reaction R_B can be computed by summing moments in the right-hand free-body diagram. Once R_B is known, reaction R_A can once again be computed by summing vertical forces, and reaction moment M_A can be obtained by summing moments about point A. Results are summarized in Fig. 3.6. Note that reaction H_A is unchanged from that reported above for the original cantilever beam structure in Fig. 3.2 (b).

$$R_B = \frac{\frac{1}{2} q_1 b\left(L - a - \frac{2}{3} b\right) + \frac{1}{2} q_2 b\left(L - a - \frac{b}{3}\right)}{L - a} \tag{3.4}$$

$$R_A = \frac{12}{13} P_3 + \left(\frac{q_1 + q_2}{2}\right)(b) - R_B \tag{3.5}$$

$$R_A = \frac{1}{78} \frac{-72 P_3 L + 72 P_3 a - 26 q_1 b^2 - 13 q_2 b^2}{-L + a} \tag{3.6}$$

$$M_A = \frac{12}{13}P_3 a + q_1 \frac{b}{2}\left(L - \frac{2}{3}b\right) + q_2 \frac{b}{2}\left(L - \frac{b}{3}\right) - R_B L \qquad (3.7)$$

$$M_A = \frac{1}{78} a \frac{-72 P_3 L + 72 P_3 a - 26 q_1 b^2 - 13 q_2 b^2}{-L + a} \qquad (3.8)$$

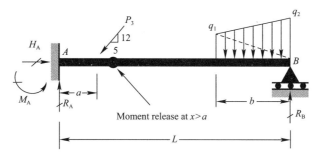

Fig. 3.6 Propped cantilever beam with moment release

The **beam with an overhang** (Fig. 3.7) supports a vertical force P_A and a couple of moment M_1. Since there are no horizontal forces acting on the beam, the horizontal reaction at the pin support is nonexistent and we do not need to show it on the free-body diagram. In arriving at this conclusion, we made use of the equation of equilibrium for forces in the horizontal direction. Consequently, only two independent equations of equilibrium remain——either two moment equations or one moment equation plus the equation for vertical equilibrium.

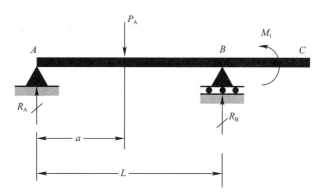

Fig. 3.7 Beam with an overhang (Repeated)

Let us **arbitrarily** decide to write two moment equations, the first for moments about point B and the second for moments about point A, as follows (counterclockwise moments are positive):

$$\sum M_B = 0 \qquad -R_A L + P_A (L-a) + M_1 = 0 \qquad (3.9)$$
$$\sum M_A = 0 \qquad -P_A a + R_B L + M_1 = 0 \qquad (3.10)$$

Therefore, the reactions are

$$R_A = \frac{P_A(L-a)}{L} + \frac{M_1}{L} \qquad R_B = \frac{P_A a}{L} - \frac{M_1}{L} \qquad (3.11)$$

Again, summation of forces in the vertical direction provides a check on these results. If the beam structure with an overhang in Fig. 3.7 (c) is modified to add a roller support at C, it is now a one degree statically indeterminate two span beam.

The stress resultants in statically determinate beams can be calculated from equations of equilibrium. In the case of the cantilever beam of Fig. 3.8 (a), we use the free-body diagram of Fig. 3.8 (b). Summing forces in the vertical direction and also taking moments about the cut section, we get

$$\sum F_{\text{vert}} = 0 \quad P - V = 0 \quad \text{or} \quad V = P \tag{3.12}$$
$$\sum M = 0 \quad M - Px = 0 \quad \text{or} \quad M = Px \tag{3.13}$$

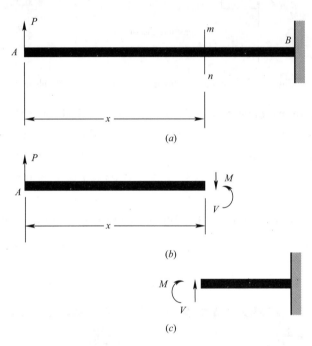

Fig. 3.8 The cantilever beam (Repeated)

where x is the distance from the free end of the beam to the cross section where V and M are being determined. Thus, through the use of a freebody diagram and two equations of equilibrium, we can calculate the shear force and bending moment without difficulty.

Let us now consider the sign conventions for shear forces and bending moments. It is customary to assume that shear forces and bending moments are positive when they act in the directions shown in Fig. 3.8 (b). Note that the shear force tends to rotate the material clockwise and the bending moment tends to compress the upper part of the beam and elongate the lower part. Also, in this instance, the shear force acts downward and the bending moment acts counterclockwise.

The action of these *same* stress resultants against the right-hand part of the beam is shown in Fig. 3.8 (c). The directions of both quantities are now reversed——the shear force acts upward and the bending moment acts clockwise. However, the shear force still tends to rotate the material clockwise and the bending moment still tends to compress the upper part of the beam and elongate the lower part.

Therefore, we must recognize that the **algebraic** sign of a stress resultant is determined by how it deforms the material on which it acts, rather than by its direction in space. In the case of a beam, a positive shear force acts clockwise against the material (Fig. 3.8b and c) and a negative shear force acts counterclockwise against the material. Also, a positive bending moment compresses the upper part of the beam (Fig. 3.8b and c) and a negative bending moment compresses the lower part.

To make these conventions clear, both positive and negative shear forces and bending moments are shown in Fig. 3.9. The forces and moments are shown acting on an element of a beam cut out between two cross sections that are a small distance apart.

Fig. 3.9 Sign convientions for shear force V and bending moment M

The deformations of an element caused by both positive and negative shear forces and bending moments are sketched in Fig. 3.10. We see that a positive shear force tends to deform the element by causing the right-hand face to move downward with respect to the left-hand face, and, as already mentioned, a positive bending moment compresses the upper part of a beam and elongates the lower part.

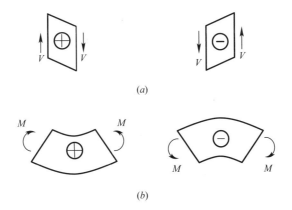

Fig. 3.10 Deformations (highly exaggerated) of a beam element caused by (a) shear forces, and (b) bending moments

Sign conventions for stress resultants are called **deformation sign conventions** because they are based upon how the material is deformed. For instance, we previously used a deformation sign convention in dealing with axial forces in a bar. We stated that an axial force producing elongation.

New Words and Expressions
[1] concentrated load 集中载荷;
[2] distributed load 分布载荷;

[3] intensity [ɪn'tensətɪ] n. 强烈；强度；烈度；
[4] uniformly distributed load 均匀分布载荷；
[5] uniform load 均布荷载；
[6] linearly varying load 线性变化荷载，线性变化载荷；
[7] couple ['kʌpl] n. 对，双；vt. & vi. 连在一起，连接；
[8] girder ['gɔːdə] n. 大梁；
[9] hinge [hɪndʒ] n. 铰链，折叶；关键，转折点；vt. & vi. 用铰链连接；
[10] bracket ['brækɪt] n. 支架，悬臂；类别，等级；vt. 为……装托架；
[11] two-dimensional ['tuːdɪ'menʃənəl] adj. 两维的；
[12] trapezoidal [træpɪ'zɔɪdəl] adj. 梯形的；
[13] centroid ['sentrɔɪd] n. 质心；矩心；面心；
[14] algebraic [ˌældʒɪ'breɪk] adj. 代数的，关于代数学的；
[15] deformation sign conventions 变形符号规则；变形符号约定。

Text C Changes in Lengths under Nonuniform Conditions

When a **prismatic** bar of linearly elastic material is loaded only at the ends, we can obtain its change in length from the equation $\delta = PL/EA$, as described in the preceding section. In this section we will see how this same equation can be used in more general situations.

Suppose, for instance, that a prismatic bar is loaded by one or more axial loads acting at intermediate points along the axis (Fig. 3.11a). We can determine the change in length of this bar by adding **algebraically** the elongations and shortenings of the individual segments. The procedure is as follows.

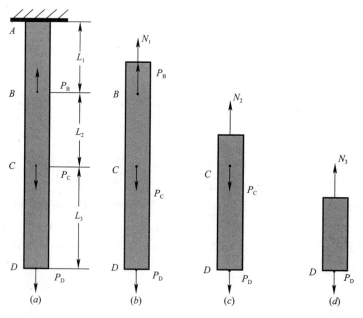

Fig. 3.11 (a) Bar with external loads acting at intermediate points; (b), (c), and (d) free-body diagrams showing the internal axial forces N_1, N_2, and N_3

1. Identify the segments of the bar (segments AB, BC, and CD) as segments 1, 2, and 3, respectively.

2. Determine the internal axial forces N_1, N_2, and N_3 in segments 1, 2, and 3, respectively, from the free-body diagrams of Fig. 3.11 (b), (c), and (d). Note that the internal axial forces are denoted by the letter N to distinguish them from the external loads P. By summing forces in the vertical direction, we obtain the following expressions for the axial forces

$$N_1 = -P_B + P_C + P_D \quad N_2 = P_C + P_D \quad N_3 = P_D \tag{3.14}$$

In writing these equations we used the sign convention given in the preceding section (internal axial forces are positive when in tension and negative when in compression).

3. Determine the changes in the lengths of the segments from Eq. (3.14)

$$\delta_1 = \frac{N_1 L_1}{EA} \quad \delta_2 = \frac{N_2 L_2}{EA} \quad \delta_3 = \frac{N_3 L_3}{EA} \tag{3.15}$$

in which L_1, L_2, and L_3 are the lengths of the segments and EA is the axial rigidity of the bar.

4. Add δ_1, δ_2 and δ_3 to obtain δ, the change in length of the entire bar

$$\delta = \sum_{i=1}^{3} \delta_i = \delta_1 + \delta_2 + \delta_3 \tag{3.16}$$

As already explained, the changes in lengths must be added algebraically, with elongations being positive and shortenings negative.

This same general approach can be used when the bar consists of several prismatic segments, each having different axial forces, different dimensions, and different materials (Fig. 3.12). The change in length may be obtained from the equation

$$\delta = \sum_{i=1}^{n} \frac{N_i L_i}{E_i A_i} \tag{3.17}$$

in which the subscript i is a numbering index for the various segments of the bar and n is the total number of segments. Note especially that N_i is not an external load but is the internal axial force in segment i.

Sometimes the axial force N and the cross-sectional area A vary continuously along the axis of a bar, as illustrated by the tapered bar of Fig. 3.11 (a). This bar not only has a continuously varying cross-sectional area but also a continuously varying axial force. In this illustration, the load consists of two parts, a single force P_B acting at end B of the bar and distributed forces $p(x)$ acting along the axis. (A distributed force has units of force per unit distance, such as pounds per inch or newtons per meter.) A distributed axial

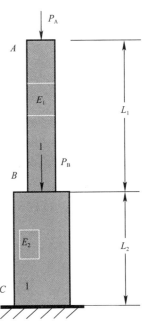

Fig. 3.12 Bar consisting of prismatic segments having different axial forces, different dimensions, and different materials

load may be produced by such factors as centrifugal forces, friction forces, or the weight of a bar hanging in a vertical position.

Under these conditions we can no longer use Eq. (3.17) to obtain the change in length. Instead, we must determine the change in length of a differential element of the bar and then integrate over the length of the bar.

We select a differential element at distance x from the left-hand end of the bar (Fig. 3.13a). The internal axial force $N(x)$ acting at this cross section (Fig. 3.13b) may be determined from equilibrium using either segment AC or segment CB as a free body. In general, this force is a function of x. Also, knowing the dimensions of the bar, we can express the cross-sectional area $A(x)$ as a function of x.

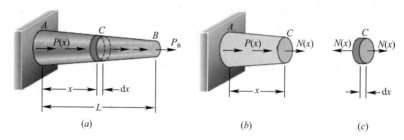

Fig. 3.13 Bar with varying cross-sectional area and vary axial force

The elongation $d\delta$ of the differential element (Fig. 3.13c) may be obtained from the equation $\delta = PL/EA$ by substituting $N(x)$ for P, dx for L, and $A(x)$ for A, as follows

$$d\delta = \frac{N(x)\,dx}{EA(x)} \qquad (3.18)$$

The elongation of the entire bar is obtained by **integrating** over the length

$$\delta = \int_0^L d\delta = \int_0^L \frac{N(x)\,dx}{EA(x)} \qquad (3.19)$$

If the expressions for $N(x)$ and $A(x)$ are not too complicated, the integral can be evaluated analytically and a formula for d can be obtained. However, if formal integration is either difficult or impossible, a numerical method for evaluating the integral should be used.

Eq. (3.17) and Eq. (3.19) apply only to bars made of linearly elastic materials, as shown by the presence of the **modulus** of elasticity E in the formulas. Also, the formula $\delta = PL/EA$ was derived using the assumption that the stress distribution is uniform over every cross section (because it is based on the formula $\sigma = P/A$). This assumption is valid for prismatic bars but not for tapered bars, and therefore Eq. (3.19) gives satisfactory results for a tapered bar only if the angle between the sides of the bar is small.

As an illustration, if the angle between the sides of a bar is 20°, the stress calculated from the expression $\sigma = P/A$ (at an arbitrarily selected cross section) is 3% less than the exact stress for that same cross section (calculated by more advanced methods). For smaller angles, the error is even less. Consequently, we can say that Eq. (3.19) is satisfactory if the angle of taper is small. If the taper is large, more accurate methods of analysis are needed.

New Words and Expressions

[1] prismatic [prɪz'mætɪk] *adj.* 棱镜的；棱形；
[2] algebraically [ældʒəb'reɪklɪ] *adv.* 用代数方法；
[3] integrating ['ɪntɪgreɪtɪŋ] *v.* 使一体化（integrate 的现在分词）；
[4] modulus ['mɒdjʊləs] *n.* 系数，模数。

Text D The Determination of Changes in Lengths of Nonuniform Bars

The springs, bars, and cables that we discussed in the preceding sections have one important feature in common——their reactions and internal forces can be determined solely from free-body diagrams and equations of equilibrium. Structures of this type are classified as **statically determinate.** We should note especially that the forces in a statically determinate structure can be found without knowing the properties of the materials. Consider, for instance, the bar AB shown in Fig. 3.14. The calculations for the internal axial forces in both parts of the bar, as well as for the reaction R at the base, are independent of the material of which the bar is made.

Most structures are more complex than the bar of Fig. 3.14, and their reactions and internal forces cannot be found by statics alone. This situation is illustrated in Fig. 3.15, which shows a bar AB fixed at both ends. There are now two vertical reactions (R_A and R_B) but only one useful equation of equilibrium——the equation for summing forces in the vertical direction. Since this equation contains two unknowns, it is not sufficient for finding the reactions. Structures of this kind are classified as **statically indeterminate.** To analyze

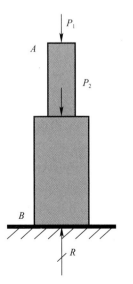
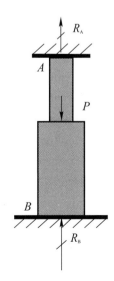

Fig. 3.14 Statically determinate bar   Fig. 3.15 Statically indeterminate bar

such structures we must supplement the equilibrium equations with additional equations pertaining to the displacements of the structure.

To see how a statically indeterminate structure is analyzed, consider the example of Fig. 3.16 (a). The prismatic bar AB is attached to rigid supports at both ends and is axially loaded by a force P at an intermediate point C. As already discussed, the reactions R_A and R_B cannot be found by statics alone, because only one **equation of equilibrium** is available:

$$\Sigma F_{\text{vert}} = 0 \qquad R_A - P + R_B = 0 \tag{3.20}$$

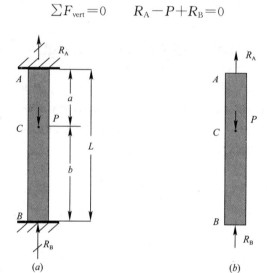

Fig. 3.16 Analysis of a statically indeterminate bar

An additional equation is needed in order to solve for the two unknown reactions. The additional equation is based upon the observation that a bar with both ends fixed does not change in length. If we separate the bar from its supports (Fig. 3.16b), we obtain a bar that is free at both ends and loaded by the three forces, R_A, R_B, and P. These forces cause the bar to change in length by an amount δ_{AB}, which must be equal to zero

$$\delta_{AB} = 0 \tag{3.21}$$

This equation, called an **equation of compatibility**, expresses the fact that the change in length of the bar must be **compatible** with the conditions at the supports.

In order to solve Eq. (3.20) and Eq. (3.21), we must now express the compatibility equation in terms of the unknown forces R_A and R_B. The relationships between the forces acting on a bar and its changes in length are known as **force-displacement relations**. These relations have various forms depending upon the properties of the material. If the material is linearly elastic, the equation $\delta = PL/EA$ can be used to obtain the force-displacement relations.

Let us assume that the bar of Fig. 3.16 has cross-sectional area A and is made of a material with modulus E. Then the changes in lengths of the upper and lower segments of the bar are, respectively,

$$\delta_{AC} = \frac{R_A a}{EA} \quad (3.22)$$

$$\delta_{CB} = -\frac{R_B b}{EA} \quad (3.23)$$

where the minus sign indicates a shortening of the bar. Eq. (3.22) and Eq. (3.23) are the force-displacement relations.

We are now ready to solve simultaneously the three sets of equations (the equation of equilibrium, the equation of compatibility, and the force-displacement relations). In this illustration, we begin by combining the force-displacement relations with the equation of compatibility

$$\delta_{AB} = \delta_{AC} + \delta_{CB} = \frac{R_A a}{EA} - \frac{R_B b}{EA} = 0 \quad (3.24)$$

Note that this equation contains the two reactions as unknowns.

The next step is to solve simultaneously the equation of equilibrium [Eq. (3.20)] and the preceding equation [Eq. (3.24)]. The results are

$$R_A = \frac{Pb}{L} \quad (3.25)$$

$$R_B = \frac{Pa}{L} \quad (3.26)$$

With the reactions known, all other force and displacement quantities can be determined. Suppose, for instance, that we wish to find the downward displacement δ_c of point C. This displacement is equal to the elongation of segment AC

$$\delta_C = \delta_{AC} = \frac{R_A a}{EA} = \frac{Pab}{LEA} \quad (3.27)$$

Also, we can find the stresses in the two segments of the bar directly from the internal axial forces (e.g., $\sigma_{AC} = RA/A = Pb/AL$).

From the preceding discussion we see that the analysis of a statically indeterminate structure involves setting up and solving equations of equilibrium, equations of compatibility, and force-displacement relations. The equilibrium equations relate the loads acting on the structure to the unknown forces (which may be reactions or internal forces), and the compatibility equations express conditions on the displacements of the structure. The force-displacement relations are expressions that use the dimensions and properties of the structural members to relate the forces and displacements of those members. In the case of axially loaded bars that behave in a linearly elastic manner, the relations are based upon the equation $\delta = PL/EA$. Finally, all three sets of equations may be solved simultaneously for the unknown forces and displacements.

In the engineering literature, various terms are used for the conditions expressed by the equilibrium, compatibility, and force-displacement equations. The equilibrium equations are also known as *static* or **kinetic** equations; the compatibility equations are sometimes called *geometric* equations, *kinematic* equations, or equations of *consistent deforma-*

tions; and the force-displacement relations are often referred to as **constitutive** *relations* (because they deal with the *constitution*, or physical properties, of the materials).

For the relatively simple structures discussed in this chapter, the preceding method of analysis is adequate. However, more formalized approaches are needed for complicated structures. Two commonly used methods, the *flexibility method* (also called the *force method*) and the *stiffness method* (also called the *displacement method*), are described in detail in textbooks on structural analysis. Even though these methods are normally used for large and complex structures requiring the solution of hundreds and sometimes thousands of simultaneous equations, they still are based upon the concepts described previously, that is, equilibrium equations, compatibility equations, and force-displacement relations.

New Words and Expressions
[1] statically determinate 静定；
[2] statically indeterminate 超静定；
[3] equation of equilibrium 平衡方程；
[4] equation of compatibility 相容方程；
[5] compatible [kəm'pætəbl] *adj.* 兼容的，相容的；
[6] force-displacement relations 力-位移关系；
[7] kinetic [kɪ'netɪk] *adj.* 运动的；[物] 动力（学）的；
[8] constitutive [ˌkɒnstɪ'tjuːtɪv] *adj.* 构成的，制定的。

LESSON 4 CONSTRUCTION MATERIALS

Text A Structural Steels

High-strength steels are used in many civil engineering projects. New steels are generally introduced under **trademarks** by their producers, but a brief check into their composition, heat treatment, and properties will normally allow them to be related to other existing materials. Following are some working classifications that allow comparison of new products with standardized ones.

Classifications of Structural Steel

General classifications allow the currently available structural steels to be grouped into four major categories, some of which have further subcategories. The steels that rely on carbon as the main alloying element are called **structural carbon steels.** The older grades in this category were the workhorse steels of the construction industry for many years, and the newer, improved carbon steels still account for the bulk of structural tonnage.

Two subcategories can be grouped in the general classification **low-alloy carbon steels.** To develop higher strengths than ordinary carbon steels, the low-alloy steels contain moderate proportions of one or more alloying elements in addition to carbon. The **columbium-vanadium-bearing steels** are higher-yield-strength metals produced by addition of small amounts of these two elements to low-carbon steels.

Two kinds of **heat-treated steels** are on the market for construction applications. **Heat-treated carbon steels** are available in either normalized or quenched-and-tempered condition, both relying essentially on carbon alone for strengthening. **Heat-treated constructional alloy steels** are quenched-and-tempered steels containing moderate amounts of alloying elements in addition to carbon.

Another general category, **maraging steels,** consists of high-nickel alloys containing little carbon. These alloys are heat-treated to age the iron-nickel martensite. Maraging steels are unique in that they are construction-grade steels that are essentially carbon-free. They rely entirely on other alloying elements to develop their high strength. This class of steels probably represents the opening of a door to the development of a whole field of carbon-free alloys.

Chemical-content comparison of carbon and other alloying elements can be used to distinguish one structural steel from another. Most structural steels, except for the maraging steels, contain carbon in amounts between 0.10% and 0.28%. The older steels have few alloying elements and are usually classified as carbon steels. Steels containing moderate amounts of alloying elements, with less than about 2% of any one constituent element, are called low-alloy steels. Steels containing larger percentages of alloying elements, such as the 18% nickel maraging steels, are designated high-alloy steels.

Heat treatment can be used as another means of classification. The older structural carbon steels and high-strength low-alloy steels are not specially heat-treated, but their properties are controlled by the hot-rolling process. The heat-treated, constructional alloy and carbon steels rely on a quenching and tempering process for development of their high-strength properties. The ASTM A514 (American Society for Testing and Materials, ASTM) steels are heat-treated by quenching in water or oil from not less than 1650°F and then tempering at not less than 1100°F. The heat-treated carbon steels are subjected to a similar quenching and tempering sequence: austenizing, water quenching, and then tempering at temperatures between 1000°F and 1300°F. The typical heat treatment of the maraging steels involves annealing at 1500°F for 1h, air cooling to room temperature, and then aging at 900°F for 3h. The aging treatment in the maraging steels may be varied to obtain different strength levels.

Mechanical Properties of Structural Steels

The tensile properties of steel are generally determined from tension tests on small specimens or coupons in accordance with standard ASTM procedures. The behavior of steels in these tests is closely related to the behavior of structural-steel members under static loads. Because, for structural steels, the yield points and moduli of elasticity determined in tension and compression are nearly the same, compression tests are seldom necessary.

Tensile strength of structural steels generally lie between about 60 and 80 ksi for the carbon and low-alloy grades and between 105 and 135 ksi for the quenched-and-tempered alloy steels (A514). Elongation in 2in, a measure of ductility, generally exceeds 20%, except for A514 steels. Modulus of elasticity usually is close to 29000 ksi.

Typical stress-strain curves for several types of steels are shown in Fig. 4.1. The initial portion of the curves is shown to a magnified scale in Fig. 4.2. It indicates that there is an initial elastic range for the structural steels in which there is no permanent deformation on removal of the load. The modulus of elasticity E, which is given by the slope of the curves, is nearly a constant 29000 ksi for all the steels. For carbon and high-strength, low-alloy steels, the inelastic range, where strains exceed those in the elastic range, consists of two parts: Initially, a plastic range occurs in which the steels yield; that is, strain increases with no increase in stress. Then follows a strain-hardening range in which increase in strain is accompanied by a significant increase in stress.

The curves in Fig. 4.2 also show an upper and lower yield point for the carbon and high-strength, low-alloy steels. The upper yield point is the one specified in standard specifications for the steels. In contrast, the curves do not indicate a yield point for the heat-treated steels. For these steels, ASTM 370, "Mechanical Testing of Steel Products,"

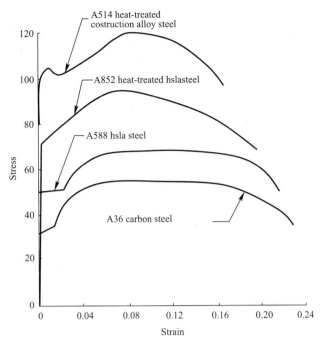

Fig. 4.1 Typical stress-strain curves for structural steels

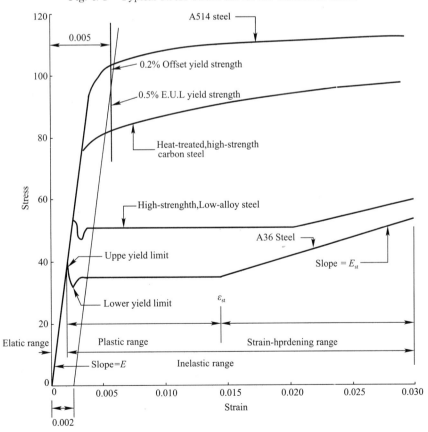

Fig. 4.2 Magnificationg of the initial portion of the stress-strain curves for structural steels shown in Fig. 4.1

recognizes two ways of indicating the stress at which there is a significant deviation from the proportionality of stress to strain. One way, applicable to steels with a specified yield point of 80 ksi or less, is to define the yield point as the stress at which a test specimen reaches a 0.5% extension under load (0.5% EUL). The second way is to define the yield strength as the stress at which a test specimen reaches a strain (offset) 0.2% greater than that for elastic behavior. Yield point and yield strength are often referred to as yield stress.

Ductility is measured in tension tests by percent elongation over a given gage length, usually 2 in or 8 in, or percent reduction of cross-sectional area. Ductility is an important property because it permits redistribution of stresses in continuous members and at points of high local stresses.

Toughness is defined as the capacity of a steel to absorb energy; the greater the capacity, the greater the toughness. Determined by the area under the stress-strain curve, toughness depends on both strength and ductility of the metal. Notch toughness is the toughness in the region of notches or other stress concentrations. A quantitative measure of notch toughness is fracture toughness, which is determined by fracture mechanics from relationships between stress and flaw size.

Poisson's ratio, the ratio of transverse to axial strain, also is measured in tension tests. It may be taken as 0.30 in the elastic range and 0.50 in the plastic range for structural steels.

Cold working of structural steels, that is, forming plates or structural shapes into other shapes at room temperature, changes several properties of the steels. The resulting strains are in the strain-hardening range. Yield strength increases but ductility decreases (Some steels are cold rolled to obtain higher strengths.). If a steel element is strained into the strain-hardening range, then unloaded and allowed to age at room or moderately elevated temperatures (a process called **strain aging**), yield and tensile strengths are increased, whereas ductility is decreased. Heat treatment can be used to modify the effects of cold working and strain aging.

Strain rate also changes the tensile properties of structural steels. In the ordinary tensile test, load is applied slowly. The resulting data are appropriate for design of structures for static loads. For design for rapid application of loads, such as impact loads, data from rapid tension tests are needed. Such tests indicate that yield and tensile strengths increase but ductility and the ratio of tensile strength to yield strength decrease.

High temperatures too affect properties of structural steels. As temperatures increase, the stress-strain curve typically becomes more rounded and tensile and yield strengths, under the action of strain aging, decrease. Poisson's ratio is not significantly affected but the modulus of elasticity decreases. Ductility is lowered until a minimum value is reached. Then, it rises with increase in temperature and becomes larger than the ductility at room temperature.

Low temperatures in combination with tensile stress and especially with geometric discontinuities, such as notches, bolt holes, and welds, may cause a brittle failure. This is a failure that occurs by cleavage, with little indication of plastic deformation. A ductile failure, in contrast, occurs mainly by shear, usually preceded by large plastic deformation. One of the most commonly used tests for rating steels on their resistance to brittle fracture is the Charpy V-notch test. It evaluates notch toughness at specific temperatures.

Hardness is used in production of steels to estimate tensile strength and to check the uniformity of tensile strength in various products. Hardness is determined as a number related to resistance to indentation. Any of several tests may be used, the resulting hardness numbers being dependent on the type of penetrator and load. These should be indicated when a hardness number is given. Commonly used hardness tests are the Brinell, Rockwell, Knoop, and Vickers.

Creep, a gradual change in strain under constant stress, is usually not significant for structural steel framing, except in fires. Creep usually occurs under high temperatures or relatively high stresses, or both.

Relaxation, a gradual decrease in load or stress under a constant strain, is a significant concern in the application of steel tendons to prestressing. With steel wire or strand, relaxation can occur at room temperature. To reduce relaxation substantially, stabilized, or low-relaxation, strand may be used. This is produced by pretensioning strand at a temperature of about 600°F. A permanent elongation of about 1% remains and yield strength increases to about 5% over stress-relieved (heat-treated but not tensioned) strand.

Residual stresses remain in structural elements after they are rolled or fabricated. They also result from uneven cooling after rolling. In a welded member, tensile residual stresses develop near the weld and compressive stresses elsewhere. Plates with rolled edges have compressive residual stresses at the edges, whereas flame-cut edges have tensile residual stresses. When loads are applied to such members, some yielding may take place where the residual stresses occur. Because of the ductility of steel, however, the effect on tensile strength is not significant but the buckling strength of columns may be lowered.

New Words and Expressions
[1] trademark ['treɪdmɑːk] *n.* 商标；
[2] classifications of structural steel 结构钢的分类；
[3] structural carbon steel 碳素结构钢；
[4] low-alloy carbon steels 低合金钢；
[5] columbium-vanadium-bearing steels 铌钒钛钢；
[6] heat-treated steels 热轧钢；
[7] heat-treated carbon steels 热轧碳素钢；
[8] heat-treated constructional alloy steels 热轧合金结构钢；
[9] maraging steels 马氏体时效钢（特高强度钢）；
[10] chemical-content comparison 化学成分含量比较；
[11] heat treatment 热处理；

［12］mechanical properties of structural steels 结构钢的力学性质；

［13］typical stress-strain curves 典型的应力-应变曲线；

［14］ductility [dʌk'tɪlɪtɪ] n. 延性；延展性；展性；可塑性；

［15］toughness ['tʌfnis] n. 强韧；韧性；韧度；

［16］poisson's ratio 泊松比；

［17］cold working 冷加工；

［18］strainrate 应变率；

［19］high temperatures 高温；

［20］low temperatures 低温；

［21］hardness ['hɑːdnəs] n. 硬度；质地；

［22］creep [kriːp] v. 蠕变；徐变；

［23］relaxation [riːlæk'seɪʃ(ə)n] n. 松弛；张弛；

［24］residual stresses 残留应力。

参考译文：结构钢材

高强度钢材在土木工程项目中得到了大量应用。钢材生产者不断用新商标将新型钢材引入土木工程领域。而对钢材的构成、热加工方式以及性能的简单了解有助于将钢材与现有的材料联系起来。如下是现阶段采用的一些可用于对比新材料与标准材料的分类方法。

结构钢分类

现有结构钢总体可以分为4大类，部分大类还可以继续细分。由碳合金构成主要元素的为碳素结构钢。一些古老等级的碳素结构钢已经在建造行业使用了多年了，而改进的新的碳素钢在现有建设中仍占有一定份额。

低合金钢这一大类可以再细分为两个小类。为了比普通的碳素钢有更高的强度，低合金钢在碳元素之外又添加了一定比例的一种或多种合金元素。铌钒钛钢是在低碳钢中添加少量铌、钒而得到的高屈服点金属。

建筑市场上有两种热轧钢材。热轧碳素钢可以分为普通型的和经过淬火-回火型的。这两种都依靠碳元素提高钢材强度。热轧低合金钢经过淬火-回火过程，除碳元素外还包含一定量的合金元素。

另一大类为马氏体时效钢（特高强度钢），包含高镍合金和少量碳元素。马氏体时效钢特别之处在于它可以是无碳的建设用钢。它们靠其他合金元素来提高强度。这种钢材为研发无碳类钢材打开了大门。

比较碳和其他元素的含量可以区分各类结构钢。马氏体时效钢之外的大部分钢材的碳含量在0.10%～0.28%。过去的钢的合金含量较少，属于碳素钢。含有一定合金元素，且任何一种合金元素占比小于2%的钢材称为低合金钢。含有较高比例的合金元素的钢材，例如含有18%镍的马氏体时效钢，称为高合金钢。

钢材还可以按照热处理方式进行分类。过去的碳素结构钢和高强度低合金钢并不经过热处理，但它们的性能也受热轧过程控制。热处理合金钢和碳素钢依靠淬火-回火过程产生高屈服点。ASTM（American Society for Testing and Materials，ASTM）A514钢材为经过如下热处理的钢材：在不超过1650°F的水或油中淬火然后回火至不超过1100°F。热处理碳素钢也经历类似淬火-回火过程：先奥氏体化，然后用水淬火，然后在 1000～1300°F 间回火。典型的马氏体时效钢需要在 1500°F 退火 1h，然后在空气中冷却至室温，然后在 900°F 时效处理 3h。对马氏体时效钢采用不同的时效处理方式可以得到不同强度等级的钢材。

结构钢的力学性质

钢材的拉伸性能通常按照标准 ASTM 流程通过对小尺寸试件的拉伸测试得到。这些测试得到的钢材性能与结构中钢构件在静力作用下的性能类似。因为结构钢的拉伸和压缩的屈服点和弹性模量基本相同，一般不需要做压缩实验。

碳素钢和低合金钢的拉伸强度通常在 60～80ksi 之间，(A514) 淬火-回火合金钢的拉伸强度在 105～135ksi 之间。伸长 2in 情况下（in 为延展性的量度），除 A514 结构钢，钢材延性通常超过 20%。弹性模量通常接近 29000ksi。

图 4.1 为几种钢材的典型应力-应变关系曲线。曲线初始阶段的比例进行了放大，如图 4.2 所示。结构钢在初始区域为弹性，即当去除荷载后结构钢没有永久变形。曲线的斜率为弹性模量 E，对所有钢材都接近 29000ksi。对于碳素钢和高强度低合金钢，应变超过了弹性范围的非弹性段包括两部分：初始部分为钢材屈服的塑性段，该段内应变增长而应力不增长。在接下来的硬化段中应变增长会伴随显著的应力增长。

图 4.2 还显示了碳素钢和高强度低合金钢具有上下两个屈服点。图中的上屈服点就是钢材规范中所定义的。而热处理钢材并没有屈服点。对于这些钢材，ASTM 370,"Mechanical Testing of Steel Products" 提供了两种方式确定应力应变比例关系变化处应力大小的方法。第一种方式适用于屈服点低于 80ksi 的钢材，当试件受拉应变达到 0.5% 即认为屈服 (0.5%EUL)。第二种方式是将屈服强度定为当应变超出弹性应变 0.2% 时的应力。屈服点、屈服强度也称为屈服应力。

延性根据拉伸试验中给定长度的伸长率确定，标距通常取为 2in 或 8in，或者根据断面收缩率确定。延性是重要的性能指标，因为它允许连续构件在高应力区域应力重分布。

韧性指的是钢材吸收能量的能力，吸收能量越大则韧性越高。韧性取决于金属的强度和延性，可以由应力-应变曲线所围成的面积确定。缺口韧性（译者：类似于我国规范的冲击韧性）指的是缺口区域或其他应力集中区域的韧性。缺口韧性可以由断裂韧性量化，断裂韧性由应力和缺陷尺寸相关的断裂力学性质确定。

泊松比指的是横向和轴向应变的比例关系，也可由拉伸试验确定。对于钢材来说，弹性段的泊松比为 0.3，塑性段的泊松比为 0.5。

钢材的冷加工是指钢材在室温加工成其他形状时钢材的一些特性会改变。冷加工产生的应变会导致应变硬化。屈服应力增加但延性降低（有些钢材经过冷轧以获得高强度）。如果钢构件进入应变硬化阶段然后再在室温或略高的温度卸载（此过程称为应变时效），屈服强度和抗拉强度都提高，但延性会降低。热处理可以减小冷加工和应变时效的影响。

应变率也会改变结构钢的拉伸性能。在标准试验中荷载是缓慢施加的。获得的结果适用于结构的静力荷载设计。如果结构受到冲击之类迅速变化的荷载，需要获得快速加载时的数据。在此类试验中，屈服强度、抗拉强度均提高，而延性和强屈比会下降。

高温也会影响结构钢的性质。当温度上升时，应力-应变曲线更圆，应变时效作用下的拉伸强度和屈服强度都降低。泊松比受高温影响较小，但弹性模量降低。延性降低至一个极小值，然后延性随着温度继续升高而升高，甚至高于室温的延性。

低温和拉应力的组合，尤其是与带有几何不连续（缺口、螺栓孔、焊缝）的构件组合时，可能出现脆性破坏。这种破坏随裂纹发生，发生前罕有塑性变形。相对而言，延性破坏大多由剪切产生，破坏前有较大的塑性变形。确定钢材抵抗脆性破坏等级的最常见的试验是夏比 V 形缺口冲击试验。它确定钢材在不同温度的缺口韧性。

在钢材生产中，硬度用于评估钢材的抗拉强度和检验不同产品抗拉强度的一致性。硬度通过抵抗压痕的能力确定。有多种试验可以确定硬度，硬度指标也与压力和压入物有关。在给定硬度时也应提供这些数据。常见的硬度测试包括 Brinell 测试、Rockwell 测试、Knoop 测试和 Vickers 测试。

蠕变指的是在持续应力状态下应变逐渐变化的现象，除火灾时，钢结构框架的蠕变通常不大。蠕变通常发生在高温或高应力状态时，或两者叠加时。

松弛是指应变保持不变，荷载或应力水平逐渐降低的现象。在预应力结构中使用钢束需要考虑松弛现象。钢丝和钢绞线的松弛现象在室温就会发生。为了从本质上降低松弛影响，可以使用稳定的或低松弛的钢绞线。这种产品是在 600°F 拉伸的。这种用热处理释放应力，不但释放拉力的钢绞线存在 1% 的永久拉伸变形，同时还可将其屈服强度提升至超过应力释放情况（热处理但未拉伸）的 5%。

构件轧制或装配后会存在残余应力。残余应力也可能是由于轧制后的不均匀降温导致的。在焊接构件中，焊缝附近产生残余拉应力，其余部位产生残余压应力。轧制钢板边缘存在残余压应力，剪切边缘存在残余拉应力。当荷载施加到此类构件时，残余应力部位可能发生屈服。但因为钢材具有延性，构件抗拉性能基本不变，但柱子的抗压能力会降低。

Text B Asphalt Binders and Asphalt Mixtures

Asphalt is one of the oldest materials used in construction. Asphalt binders were used in 3000 B.C., preceding the use of the wheel by 1000 years. Before the mid-1850s, asphalt came from natural pools found in various locations throughout the world, such as the Trinidad Lake asphalt, which is still mined. However, with the discovery and refining of petroleum in Pennsylvania, use of asphalt cement became widespread. By 1907, more asphalt cement came from refineries than came from natural deposits. Today, practically all asphalt cement is from refined petroleum.

Types of Asphalt Products

Asphalt used in pavements is produced in three forms: **asphalt cement**, **asphalt cutback**, and **asphalt emulsion**. Asphalt cement is a blend of hydrocarbons of different molecular weights. The characteristics of the asphalt depend on the chemical composition and the distribution of the molecular weight hydrocarbons. As the distribution shifts toward heavier molecular weights, the asphalt becomes harder and more viscous. At room temperatures, asphalt cement is a semisolid material that cannot be applied readily as a binder without being heated. Liquid asphalt products, cutbacks and emulsions, have been developed and can be used without heating (The Asphalt Institute 1989).

Although the liquid asphalts are convenient, they cannot produce a quality of asphalt concrete comparable to what can be produced by heating neat asphalt cement and mixing it with carefully selected aggregates. Asphalt cement has excellent adhesive characteristics, which make it a superior binder for pavement applications. In fact, it is the most common binder material used in pavements.

A cutback is produced by dissolving asphalt cement in a lighter molecular weight hydrocarbon solvent. When the cutback is sprayed on a pavement or mixed with aggregates, the solvent evaporates, leaving the asphalt residue as the binder. In the past, cutbacks were widely used for highway construction. They were effective and could be applied easily in the field. However, three disadvantages have severely limited the use of cutbacks. First, as petroleum costs have escalated, the use of these expensive solvents as a carrying agent for the asphalt cement is no longer cost effective. Second, cutbacks are hazardous materials due to the volatility of the solvents. Finally, application of the cutback releases environmentally unacceptable hydrocarbons into the atmosphere. In fact, many regions with air pollution problems have outlawed the use of any cutback material.

An alternative to dissolving the asphalt in a solvent is dispersing the asphalt in water as emulsion. In this process the asphalt cement is physically broken down into micron-sized globules that are mixed into water containing an emulsifying agent. Emulsified asphalts typically consist of about 60% to 70% asphalt residue, 30% to 40% water, and a fraction of a percent of emulsifying agent. There are many types of emulsifying agents; basically they are a soap material. The emulsifying molecule has two distinct components, the head portion, which has an electrostatic charge, and the tail portion, which has a high affinity for asphalt. The charge can be either positive to produce a cationic emulsion or negative to produce an anionic emulsion. When asphalt is introduced into the water with the emulsifying agent, the tail portion of the emulsifier attaches itself to the asphalt, leaving the head exposed. The electric charge of the emulsifier causes a repulsive force between the asphalt globules, which maintains their separation in the water. Since the specific gravity of asphalt is very near that of water, the globules have a neutral buoyancy and, therefore, do not tend to float or sink. When the emulsion is mixed with aggregates or used on a pavement, the water evaporates, allowing the asphalt globs to come together, forming the binder. The phenomenon of separation between the asphalt residue and water is referred to as breaking or setting. The rate of emulsion setting can be controlled by varying the type and amount of the emulsifying agent.

Since most aggregates bear either positive surface charges (such as limestone) or negative surface charges (such as siliceous aggregates), they tend to be compatible with **anionic or cationic emulsions**, respectively. However, some emulsion manufacturers can produce emulsions that bond well to aggregate-specific types, regardless of the surface charges.

Although emulsions and cutbacks can be used for the same applications, the use of emulsions is increasing because they do not include hazardous and costly solvents.

Uses of Asphalt

The main use of asphalt is in pavement construction and maintenance (Tab. 4.1). In addition, asphalt is used in **sealing** and waterproofing various structural components, such as roofs and underground foundations.

Tab. 4.1 Paving applications of asphalt

Term	Description	Application
Hot mix asphalt	Carefully designed mixture of asphalt and aggregates	Pavement surface, patching
Cold mix	Mixture of aggregates and liquid asphalt	Patching, low volume road surface, asphalt stabilized base
Fog seal	Spray of diluted asphalt emulsion on existing pavement surface	Seal existing pavement surface
Prime coat	Spray coat to bond aggregate base and asphalt concrete surface	Construction of flexible pavement
Tack coat	Spray coat between lifts of asphalt concrete	Construction of new pavements or between an existing pavement and an overlay
Chip seal	Spray coat of asphalt emulsion (or asphalt cement or cutback) followed with aggregate layer	Maintenance of existing pavement or low volume road surfaces
Slurry seal	Mixture of emulsion, well-graded fine aggregate and water	Resurface low volume roads
Microsurfacing	Mixture of polymer modified emulsion, well-graded crushed fine aggregate, mineral filler, water, and additives	Texturing, sealing, crack filling, rut filling, and minor leveling

The selection of the type and grade of asphalt depends on the type of construction and the climate of the area. Asphalt cements, also called asphalt binders, are used typically to make hot-mix asphalt concrete for the surface layer of asphalt pavements. Asphalt concrete is also used in patching and repairing both asphalt and portland cement concrete pavements. Liquid asphalts (emulsions and cutbacks) are used for pavement maintenance applications, such as **fog seals**, chip seals, **slurry seals**, and microsurfacing. Liquid asphalts may also be used to seal the cracks in pavements. Liquid asphalts are mixed with aggregates to produce cold mixes, as well. Cold mixtures are normally used for patching (when hot-mix asphalt concrete is not available), base and **subbase stabilization**, and surfacing of low-volume roads.

Temperature Susceptibility of Asphalt

The consistency of asphalt is greatly affected by temperature. Asphalt gets hard and **brittle** at low temperatures and soft at high temperatures. Fig. 4.3 shows a conceptual relation between temperature and logarithm of viscosity. The viscosity of the asphalt decreases when the temperature increases. Asphalt's temperature susceptibility can be represented by the slope of the line shown in Fig. 4.3. The steeper the slope the higher the temperature susceptibility of the asphalt. However, additives can be used to reduce this susceptibility.

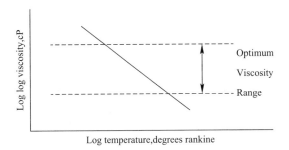

Fig. 4.3 Typical relation between asphalt viscosity and temperature

When asphalt is mixed with aggregates, the mixture will perform properly only if the asphalt viscosity is within an optimum range. If the viscosity of asphalt is higher than the optimum range, the mixture will be too brittle and susceptible to low-temperature cracking. On the other hand, if the viscosity is below the optimum range, the mixture will flow readily, resulting in permanent deformation (rutting).

Due to temperature susceptibility, the grade of the asphalt cement should be selected according to the climate of the area. The viscosity of the asphalt should be mostly within the optimum range for the area's annual temperature range; soft-grade asphalts are used for cold climates and hard-grade asphalts for hot climates (See Fig. 4.4).

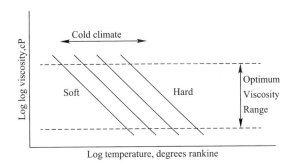

Fig. 4.4 Selecting the proper grade of asphalt binder to match the climate

Asphalt Concrete

Asphalt concrete, also known as hot-mix asphalt (HMA), consists of asphalt cement and aggregates mixed together at a high temperature and placed and compacted on the road while still hot. Asphalt (flexible) pavements cover approximately 93% of the 2 million miles of paved roads in the United States, while the remaining 7% of the roads are portland cement concrete (rigid) pavements. The performance of asphalt pavements is largely a function of the asphalt concrete surface material.

The objective of the asphalt concrete mix design process is to provide the following properties (Roberts et al. 1996):

1. stability or resistance to permanent deformation under the action of traffic loads, especially at high temperatures.

2. **fatigue** resistance to prevent fatigue cracking under repeated loadings.

3. resistance to thermal cracking that might occur due to contraction at low temperatures.

4. resistance to hardening or aging during production in the mixing plant and in service.

5. resistance to moisture-induced damage that might result in stripping of asphalt from aggregate particles.

6. skid resistance, by providing enough texture at the pavement surface.

7. workability, to reduce the effort needed during mixing, placing and compaction.

Regardless of the set of criteria used to state the objectives of the mix design process, the design of asphalt concrete mixes requires compromises. For example, extremely high stability often is obtained at the expense of lower durability, and vice versa. Thus, in evaluating and adjusting a mix design for a particular use, the aggregate gradation and asphalt content must strike a favorable balance between the stability and durability requirements. Moreover, the produced mix must be practical and economical.

New Words and Expressions

[1] asphalt ['æsfælt] *n.* 沥青；柏油路；
[2] types of asphalt products 沥青产品类型；
[3] asphalt cement 沥青水泥；
[4] cutback ['kʌtbæk] *v.* 削减；缩减；回调；
[5] asphalt emulsion 乳化沥青；乳状沥青；
[6] anionic or cationic emulsions 阴离子或阳离子乳液；
[7] sealing ['siːlɪŋ] *v.* 密封；封闭；嵌缝；
[8] fog seals 雾封层；
[9] slurry seals 沥青封层；
[10] subbase stabilization 底基稳定；
[11] temperature susceptibility of asphalt 沥青的温度敏感性；
[12] brittle ['brɪtl] *adj.* 易碎的；脆弱的；
[13] fatigue [fə'tiːg] *v.* 疲劳，疲乏。

Text C Portland Cement Concretes

This mixture of **portland cement fine aggregate, coarse aggregate**, air, and water is a temporarily plastic material, which can be cast or molded, but is later converted to a solid mass by chemical reaction. The user of concrete desires adequate strength, placeability, and durability at minimum cost. The concrete designer may vary the proportions of the five constituents of concrete over wide limits to attain these aims. The principal variables are the water-cement ratio, cement aggregate ratio, size of coarse aggregate, ratio of fine aggregate to coarse aggregate, type of cement, and use of admixtures.

Established basic relationships and laboratory tests provide guidelines for approaching optimum combinations. ACI 211.1, "Recommended Practice for Selecting Proportions for

Normal and Heavyweight Concrete", and ACI 211.2, "Recommended Practice for Selecting Proportions for Structural Lightweight Concrete".

Aggregates for Portland Cement Concretes

Aggregate is a broad term encompassing boulders, cobbles, crushed stone, gravel, air-cooled blast furnace slag, native and manufactured sands, and manufactured and natural lightweight aggregates. Aggregates may be further described by their respective sizes.

Normal-Weight Aggregates. These typically have specific gravities between 2.0 and 3.0. They are usually distinguished by size as follows:

Boulders	Larger than 6in
Cobbles	6 to 3in
Coarse aggregate	3in to No. 4 sieve
Fine aggregate	No. 4 sieve to No. 200 sieve
Mineral filler	Material passing No. 200 sieve

Used in most concrete construction, normal-weight aggregates are obtained by dredging riverbeds or mining and crushing formational material. Concrete made with normal-weight fine and coarse aggregates generally weighs about $144 lb/ft^3$.

Boulders and cobbles are generally not used in their as-mined size but are crushed to make various sizes of coarse aggregate and manufactured sand and mineral filler. Gravels and naturally occurring sand are produced by the action of water and weathering on glacial and river deposits. These materials have round, smooth surfaces and particle-size distributions that require minimal processing. These materials can be supplied in either coarse or fine-aggregate sizes.

Fine aggregates have 100% of their material passing the 3/8 in sieve. Coarse aggregates have the bulk of the material retained on the No. 4 sieve.

Aggregates comprise about 75%, by volume, of a typical concrete mix. Cleanliness, soundness, strength, and particle shape are important in any aggregate. Aggregates are considered clean if they are free of excess clay, silt, mica, organic matter, chemical salts, and coated grains. An aggregate is physically sound if it retains dimensional stability under temperature or moisture change and resists weathering without decomposition. To be considered adequate in strength, an aggregate should be able to develop the full strength of the cementing matrix. When wear resistance is important, the aggregate should be hard and tough.

Several processes have been developed for improving the quality of aggregates that do not meet desired specifications. Washing may be used to remove particle coatings or change aggregate gradation. Heavy-media separation, using a variable-specific-gravity liquid, such as a suspension of water and finely ground magnetite and ferrosilicon, can be used to improve coarse aggregates. Deleterious lightweight material is removed by flotation, and heavyweight particles settle out. Hydraulic jigging, where lighter particles are carried upward by pulsations caused by air or rubber diaphragms, is also a means for separation of lighter particles. Soft, friable particles can be separated from hard, elastic particles by a process called elastic fractionation. Aggregates are dropped onto an inclined hardened-steel surface, and their quality is measured by the distance they bounce.

Aggregates that contain certain forms of silicas or carbonates may react with the alkalies present in portland cement (sodium oxide and potassium oxide). The reaction product cracks the concrete or may create pop-outs at the concrete surface. The reaction is more pronounced when the concrete is in a warm, damp environment.

The potential reactivity of an aggregate with alkalies can be determined either by a chemical test or by a mortar-bar method. The mortar-bar method is the more rigorous test and provides more reliable results but it requires a much longer time to perform.

Lightweight Aggregates. Lightweight aggregates are produced by expanding clay, shale, slate, perlite, obsidian, and vermiculite with heat; by expanding blast-furnace slag through special cooling processes; from natural deposits of pumice, scoria, volcanic cinders, tuff, and diatomite; and from industrial cinders. The strength of concrete made with lightweight aggregates is roughly proportional to its weight, which may vary from 35 to 115 lb/ft^3.

Lightweight aggregates can be divided into two categories: structural and nonstructural. The structural lightweight aggregates are either manufactured (expanded clay, shale, or slate, or blast-furnace slag) or natural (scoria and pumice). These aggregates produce concretes generally in the strength range of 3000 to 4000 psi; higher strengths are attainable.

The common nonstructural lightweight aggregates are vermiculite and perlite, although scoria and pumice can also be used. These materials are used in insulating concrete for soundproofing and nonstructural floor toppings. Lightweight concrete has better fire resistance and heat-and sound-insulation properties than ordinary concrete, and it offers savings in structural supports and decreased foundations due to decreased dead loads. Structural concrete with lightweight aggregates costs 30% to 50% more, however, than that made with ordinary aggregates and has greater porosity and more drying shrinkage. Resistance to weathering is about the same for both types of concrete. Lightweight concrete can also be made with foaming agents, such as aluminum powder, which generates a gas while the concrete is still plastic and may be expanded.

Heavy Aggregates. In the construction of atomic reactors, large amounts of heavyweight concrete are used for shielding and structural purposes. Heavy aggregates are used in shielding concrete because gamma-ray absorption is proportional to density. Heavy concrete may vary between the 150 lb/ft^3 weight of conventional sand-and-gravel concrete and the theoretical maximum of 384 lb/ft^3 where steel shot is used as fine aggregate and steel punchings as coarse aggregate. In addition to manufactured aggregates from iron products, various quarry products and ores, such as barite, limonite, hematite, ilmenite, and magnetite, have been used as heavy aggregates.

Normal-Weight Concrete

The nominal weight of normal concrete is 144lb/ft^3 for non-air-entrained concrete but is less for air-entrained concrete (The weight of concrete plus steel reinforcement is often assumed as 150lb/ft^3).

Strength for normal-weight concrete ranges from 2000 to 20000 psi. It is generally measured using a standard test cylinder 6in in diameter by 12in high. The strength of a concrete is defined as the average strength of two cylinders taken from the same load and tested at the same age. Flexural beams 6in×6in×20in may be used for concrete paving mixes.

Water-Cement (W/C) ratio is the prime factor affecting the strength of concrete. Fig. 4.5 shows how W/C, expressed as a ratio by weight, affects the compressive strength for both air-entrained and non-air-entrained concrete. Strength decreases with an increase in W/C in both cases.

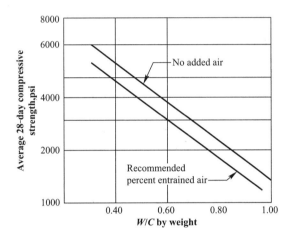

Fig. 4.5 Concrete strength decreases with increase in water-cement ratio for concrete with and without entrained air (From "Concrete Manual," 8th ed., U. S. Bureau of Reclamation.)

Cement content itself affects the strength of concrete, with strength decreasing as cement content is decreased. In air-entrained concrete, this strength decrease can be partly overcome by taking advantage of the increased workability due to air entrainment, which permits a reduction in the amount of water.

Because of the water reduction possibility, the strengths of air-entrained concrete do not fall as far below those for non-air-entrained concrete as those previously indicated in Fig. 4.5.

Curing conditions are vital in the development of concrete strength. Since cement-hydration reactions proceed only in the presence of an adequate amount of water, moisture must be maintained in the concrete during the curing period. Curing temperature also affects concrete strength. Longer periods of moist curing are required at lower temperatures to develop a given strength. Although continued curing at elevated temperatures results in faster strength development up to 28 days, at later ages the trend is reversed; concrete cured at lower temperatures develops higher strengths.

Note that concrete can be frozen and will not gain strength in this state. Note also that, at low temperatures, strength gain of non-frozen concrete is minimal and environmental factors, especially temperature and curing, are extremely important in development of concrete strength.

Stress-Strain Relations. Concrete is not a linearly elastic material; the stress-strain relation for continuously increasing loading plots as a curved line. For concrete that has hardened thoroughly and has been moderately preloaded, however, the stress-strain curve is practically a straight line within the range of usual working stresses. As shown in Fig. 4.6, a modulus of elasticity can be determined from this portion of the curve. The elastic modulus for ordinary concretes at 28 days ranges from 2000 to 6000ksi.

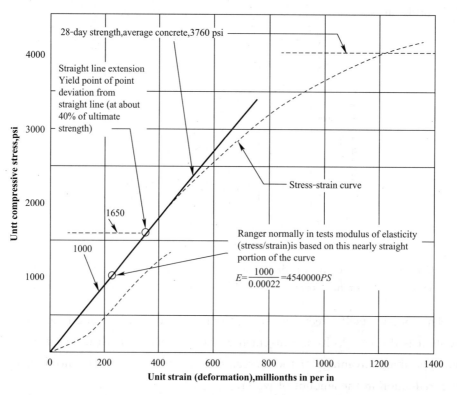

Fig. 4.6 Typical stress-strain diagram for cured concrete that has been moderately preloaded (From "Concrete Manual," 8th ed., U. S. Bureau of Reclamation.)

In addition to the elastic deformation that results immediately upon application of a load to concrete, deformation continues to increase with time under a sustained load. This plastic flow, or creep, continues for an indefinite time. It proceeds at a continuously diminishing rate and approaches a limiting value which may be one to three times the initial elastic deformation. Although increasing creep-deformation measurements have been recorded for periods in excess of 10 years, more than half of the ultimate creep usually takes place within the first 3 months after loading. Upon unloading, an immediate elastic recovery takes place, followed by a plastic recovery of lesser amount than the creep on the first loading.

Volume changes play an important part in the durability of concrete. Excessive or differential volume changes can cause cracking as a result of shrinkage and insufficient tensile strength, or spalling at joints due to expansion. Swelling and shrinkage of concrete occur with changes in moisture within the cement paste.

Hardened cement paste contains minute pores of molecular dimensions between particles of tobermorite gel and larger pores between aggregations of gel particles. The volume of pore space in a cement paste depends on the initial amount of water mixed with the cement; any excess water gives rise to additional pores, which weaken the structure of the cement paste. Movements of moisture into and out of this pore system cause volume changes. The drying shrinkage of concrete is about $\frac{1}{2}$ in/100ft. There is a direct relationship between mix-water content and drying shrinkage. The cement content is of secondary importance in shrinkage considerations.

The thermal coefficient of expansion of concrete varies mainly with the type and amount of coarse aggregate used. The cement paste has a minor effect. An average value used for estimating is 5.5×10^{-6} in/(in°F).

New Words and Expressions
[1] portland cement fine aggregate 波特兰水泥细骨料；
[2] coarse aggregate 粗骨料；
[3] normal-weight aggregates 标准重量的骨料；
[4] lightweight aggregates 轻质骨料；
[5] heavy aggregates 高强骨料；
[6] water-cement ratio 水灰比；
[7] curing conditions 养护条件；
[8] stress-strain relations 应力-应变关系；
[9] volume changes 体应变；
[10] the thermal coefficient of expansion 热膨胀系数。

LESSON 5 DESIGN OF CONCRETE STRUCTURE

Text A Design of Concrete Beams

Ultimate-Strength Theory for Reinforced Concrete Beams

For consistent, safe, economical design of beams, their actual load-carrying capacity should be known. The safe load then can be determined by dividing this capacity by a safety factor. Or the design load can be multiplied by the safety factor to indicate what the capacity of the beams should be. It should be noted, however, that under service loads, stresses and deflections may be computed with good approximation on the assumption of a linear stress-strain diagram and a cracked cross section.

ACI 318, "Building Code Requirements for Reinforced Concrete" (American Concrete Institute), provides for design by **ultimate-strength theory. Bending moments** in members are determined as if the structure were elastic. Ultimate-strength theory is used to design critical sections, those with the largest bending moments, shear, **torsion,** etc. The ultimate strength of each section is computed, and the section is designed for this capacity.

Stress Redistribution

The ACI Code recognizes that, below ultimate load, a redistribution of stress occurs in continuous beams, frames, and arches. This allows the structure to carry loads higher than those indicated by elastic analysis. The code permits an increase or decrease of up to 10% in the negative moments calculated by elastic theory at the supports of continuous flexural members. But these modified moments must also be used for determining the moments at other sections for the same loading conditions. [The modifications, however, are permissible only for relatively small **steel ratios** at each support. The steel ratios ρ or $\rho - \rho'$ should be less than half ρ_b, the steel ratio for balanced conditions (concrete strength equal to steel strength) at ultimate load.] For example, suppose elastic analysis of a continuous beam indicates a maximum negative moment at a support of $wL^2/12$ and maximum positive moment at mid-span of $wL^2/8 \sim wL^2/12$, or $wL^2/24$. Then, the code permits the negative moment to be decreased to $0.9wL^2/12$, if the positive moment is increased to $wL^2/8 \sim 0.9wL^2/12$, or $1.2wL^2/24$.

Design Assumptions for Ultimate-Strength Design

Ultimate strength of any section of a reinforced-concrete beam may be computed assuming the following:

1. Strain in the concrete is directly proportional to the distance from the neutral axis

(Fig. 5.1b).

2. Except in **anchorage zones**, strain in reinforcing steel equals strain in adjoining concrete.

3. At ultimate strength, maximum strain at the extreme compression surface equals 0.003 in/in.

4. When the reinforcing steel is not stressed to its yield strength f_y, the steel stress is 29000 ksi times the steel strain, in/in. After the yield strength has been reached, the stress remains constant at f_y, though the strain increases.

5. Tensile strength of the concrete is negligible.

At ultimate strength, concrete stress is not proportional to strain. The actual stress distribution may be represented by an equivalent rectangle, known as the whitney rectangular stress block, that yields ultimate strengths in agreement with numerous, comprehensive tests (Fig. 5.1c).

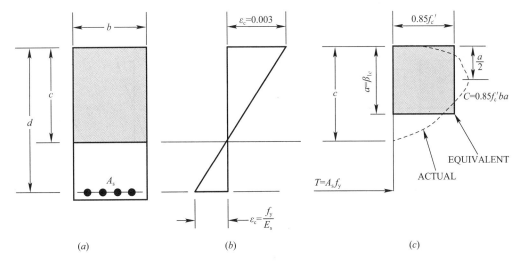

Fig. 5.1 Stresses and strains on a reinforced-concrete beam section
(a) At ultimate load, after the section has crackde and only the steel carries tension;
(b) Strain diagram; (c) Actual and assumed compression-stress

The ACI Code recommends that the compressive stress for the equivalent rectangle be taken as $0.85f'_c$, where f'_c is the 28-day compressive strength of the concrete. The stress is assumed constant from the surface of maximum compressive strain over a depth $a = \beta_1 c$, where c is the distance to the neutral axis (Fig. 5.1c). For $f'_c \leqslant 4000\text{psi}$, $\beta_1 = 0.85$; for greater concrete strengths, β_1 is reduced 0.05 for each 1000psi in excess of 4000psi.

Formulas in the ACI Code based on these assumptions usually contain a factor f which is applied to the theoretical ultimate strength of a section, to provide for the possibility that small adverse variations in materials, quality of work, and dimensions, while individually within acceptable tolerances, occasionally may combine, and actual capacity may be less than that computed. The coefficient φ is taken as 0.90 for **flexure**, 0.85 for shear and torsion, 0.75 for spirally reinforced compression members, and 0.70 for tied compression members. Under certain conditions of load (as the value of the axial load approaches zero) and geometry, the φ value for compression members may increase linearly to a maximum value of 0.90.

Working-Stress Theory for Reinforced-Concrete Beams

Stress distribution in a reinforced-concrete beam under service loads is different from that at ultimate strength. Knowledge of this stress distribution is desirable for many reasons, including the requirements of some design codes that specified working stresses in steel and concrete not be exceeded.

Working stresses in reinforced-concrete beams are computed from the following assumptions:

1. **Longitudinal stresses and strains** vary with distance from the neutral axis (Fig. 5.2c and d); that is, plane sections remain plane after bending. (Strains in longitudinal reinforcing steel and adjoining concrete are equal.)

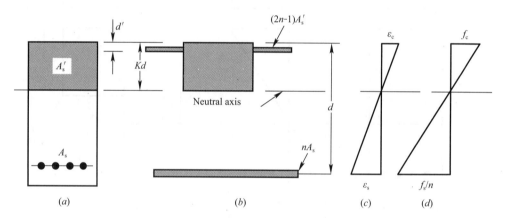

Fig. 5.2 Typical cracked cross section of a reinforced concrete beam
(a) Only the reinforcing steel is effective in tension; (b) Section treated as an all-concrete transformed section. In working-stress design, inear distributon is assumed for (c) strains and (d) stresses

2. The concrete does not develop any tension. (Concrete cracks under tension.)

3. Except in anchorage zones, strain in reinforcing steel equals strain in adjoining concrete. But because of creep, strain in compressive steel in beams may be taken as half that in the adjoining concrete.

4. The modular ratio $n = E_s/E_c$ is constant. E_s is the modulus of elasticity of the reinforcing steel and E_c of the concrete.

Other **equivalency** factors are also given in terms of ultimate-strength values. Thus, the predominant design procedure is the ultimate-strength method, but for reasons of background and historical significance and because the working-stress design method is sometimes preferred for bridges and certain foundation and retaining-wall design.

Transformed Section. According to the **working-stress** theory for reinforced-concrete beams, strains in reinforcing steel and adjoining concrete are equal. Hence f_s, the stress in the steel, is n times f_c, the stress in the concrete, where n is the ratio of modulus of elasticity of the steel E_s to that of the concrete E_c. The total force acting on the steel then equals $(nA_s) f_c$. This indicates that the steel area can be replaced in **stress calculations** by a concrete area n times as large.

The transformed section of a concrete beam is one in which the reinforcing has been replaced by an equivalent area of concrete (Fig. 5.2b). (In doubly reinforced beams and **slabs**, an effective modular ratio of $2n$ should be used to transform the **compression reinforcement**, to account for the effects of creep and nonlinearity of the stress-strain diagram for concrete. But the computed stress should not exceed the allowable tensile stress.) Since stresses and strains are assumed to vary with distance from the neutral axis, conventional elastic theory for homogeneous beams holds for the transformed section. Section properties, such as location of neutral axis, **moment of inertia**, and section modulus S, can be computed in the usual way, and stresses can be found from the flexure formula $f=M/S$, where M is the bending moment.

Deflection Computations and Criteria for Concrete Beams

The assumptions of working-stress theory may also be used for computing deflections under service loads; that is, elastic-theory deflection formulas may be used for reinforced-concrete beams. In these formulas, the effective moment of inertia I_e is given by Eq. (5.1)

$$I_e = \left(\frac{M_{cr}}{M_a}\right)^3 I_g + \left[1 - \left(\frac{M_{cr}}{M_a}\right)^3\right] I_{cr} \leqslant I_g \tag{5.1}$$

where I_g——moment of inertia of the gross concrete section;
   M_{cr}——cracking moment;
   M_a——moment for which deflection is being computed;
   I_{cr}——cracked concrete (transformed) section.

If y_t is taken as the distance from the **centroidal axis** of the gross section, neglecting the reinforcement, to the extreme surface in tension, the cracking moment may be computed from

$$M_{cr} = \frac{f_r I_g}{y_t} \tag{5.2}$$

with the **modulus of rupture of the concrete** $f_r = 7.5\sqrt{f_c'}$. Eq. (5.1) takes into account the variation of the moment of inertia of a concrete section based on whether the section is cracked or uncracked. The modulus of elasticity of the concrete E_c may be computed.

The deflections thus calculated are those assumed to occur immediately on application of load. The total long-term deflection is

$$\Delta_{LT} = \Delta_L + \lambda_\infty \Delta_D + \lambda_t \Delta_{LS} \tag{5.3}$$

where Δ_L——initial live load deflection;
   Δ_D——initial dead load deflection;
   Δ_{LS}——initial sustained live-load deflection;
   λ_∞——time dependent multiplier for infinite duration of sustained load;
   λ_t——time dependent multiplier for limited load duration.

Deflection Limitations. The ACI Code recommends the following limits on deflections in buildings:

For roofs not supporting and not attached to nonstructural elements likely to be damaged by large deflections, maximum immediate deflection under live load should not exceed $L/180$, where L is the span of beam or slab.

For floors not supporting partitions and not attached to nonstructural elements, the maximum immediate deflection under live load should not exceed $L/360$.

For a floor or roof construction intended to support or to be attached to partitions or other construction likely to be damaged by large deflections of the support, the allowable limit for the sum of immediate deflection due to live loads and the additional deflection due to shrinkage and creep under all sustained loads should not exceed $L/480$. If the construction is not likely to be damaged by large deflections, the deflection limitation may be increased to $L/240$. But tolerances should be established and adequate measures should be taken to prevent damage to supported or nonstructural elements resulting from the deflections of structural members.

New Words and Expressions

[1] ultimate-strength theory 极限强度理论；
[2] bending moments 弯矩；
[3] torsion ['tɔːʃn] n. 扭转；
[4] stress redistribution 应力重分布；
[5] steel ratios 钢筋比；
[6] anchorage zones 锚固区；
[7] flexure ['flekʃə] n. 屈曲；
[8] longitudinal stresses and strains 纵向应力和应变；
[9] equivalency [ɪ'kwɪvələns] n. 等效；
[10] transformed section 换算截面；
[11] working-stress 工作应力；
[12] stress calculations 应力计算；
[13] slab [slæb] n. 板；
[14] compression reinforcement 受压钢筋；
[15] moment of inertia 转动惯量；
[16] centroidal axis 形心轴；
[17] modulus of rupture of the concrete 混凝土的断裂模数；
[18] deflection limitations 偏转限制。

参考译文：混凝土梁的设计

钢筋混凝土梁的极限强度理论

为了保证梁的设计能持久、安全、经济，需要知道梁的实际承载能力。然后将实际承载能力除以一个安全系数可以得到安全负荷。或者，将设计荷载乘以一个安全系数来确定梁的承载能力。然而需要指出的是，在使用荷载下，这些应力和变形可以用线性的应力-应变图和裂纹截面的概念进行很好的近似计算。

ACI 318 规范（美国混凝土协会颁布的《建筑钢筋混凝土结构规范》）为按极限强度理论提供的设计规范。构件中的弯矩按照结构为弹性来确定的。极限强度理论用于设计关键截面，即弯矩、剪切、扭转等最大的截面。计算各个截面的极限强度后，这个截面就是按这个承载能力而设计的。

应力重分布

ACI 规范指出，在极限荷载下连续梁、框架和拱上会发生应力重分布。这使得结构能够承受比弹性分析更高的荷载。规范允许在连续的受弯构件的支座处，按弹性理论计算的负弯矩最多增减10%。但计算同种工况下其他位置的弯矩时也应采用这些修正后的弯矩[这种修改只允许在每个支撑点配筋率相对较小时进行。配筋率 ρ 或 $\rho - \rho'$ 应小于 ρ_b 的一半，ρ_b 指极限荷载下按平衡条件（混凝土强度等于钢的强度）的钢筋比率]。例如，假设一个连续梁按弹性分析得到的支座最大负弯矩 $wL^2/12$，跨中最大正弯矩 $wL^2/8 \sim wL^2/12$ 或 $wL^2/24$。如果正弯矩被增加到 $wL^2/8 \sim 0.9wL^2/12$，或者 $1.2wL^2/24$，则规范允许负弯矩应该降低到 $0.9wL^2/12$。

极限强度设计假设

任一钢筋混凝土桥梁的截面的极限应力的计算基于如下假设：

1. 混凝土中的应变与离中性轴的距离成正比（图5.1b）。
2. 除锚固区外，钢筋的应变和相邻混凝土的应变相等。
3. 在极限强度下，极限压缩面最大应变等于0.003。
4. 当配筋强度未达到屈服强度 f_y 时，钢的应力为钢应变乘以29000ksi，在达到屈服强度后，应力在 f_y 处保持不变，但应变增大。
5. 混凝土的抗拉强度可以忽略不计。

达到极限强度时混凝土的应力与应变不成正比。实际的应力分布可以用一个等效的矩形来表示，称为惠特尼矩形应力块，它的极限屈服强度和大量综合测试结果一致(图5.1c)。

ACI 规范建议等效矩形的压应力应该取为 $0.85f'_c$，其中 f'_c 为混凝土28d的抗压强度。假设从表面到深度 $a = \beta_1 c$ 的应力为一常数，其中 c 是到中性轴的距离（图5.1c）。当 $f'_c \leqslant$ 4000psi 时，$\beta_1 = 0.85$；当混凝土强度更大时，从4000psi以上每增加1000psi，β_1 降低0.05。

ACI 规范是基于这些假设的，在计算截面的理论最大强度时通常使用参数 f，以考虑材料、施工质量、尺寸偏差等微小的不利变化。这些不利变化单独可能在允许范围内，但偶尔耦合起来会使得实际的承载力低于计算的承载力。对于弯曲系数 φ 取 0.9，剪切和扭转取 0.85，螺旋形的钢混受压构件取 0.75，普通箍筋受压构件取 0.7。在一定的荷载条件下（当轴向荷载接近于0时）和几何形状下，受压构件的 φ 值可以线性增大到最大值0.9。

钢筋混凝土梁的工作应力理论

钢筋混凝土梁在工作荷载作用下的应力分布与极限荷载作用下的分布不同。了解这种应力分布是必要的，因为一些设计规范要求钢和混凝土不应超过特定的工作应力。

计算钢筋混凝土梁的工作应力基于以下假设：

1. 纵向应力和应变随距中性轴距离而变化（图5.2c和d），即弯曲后的平面截面仍为平面（纵向钢筋相邻的混凝土应变相等）。

2. 混凝土不会产生任何张力（混凝土在拉力的作用下开裂）。

3. 除锚固区外，钢筋的应变与相邻的混凝土应变相等。由于徐变的原因，梁内受压部分钢的应变可以取相邻混凝土应变的一半。

4. 模量比 $n=E_s/E_c$ 是恒定的，E_s 为钢筋的弹性模量，E_c 是混凝土的弹性模量。

根据极限强度值还给出了其他等效参数。因此主流的设计流程是极限应力方法，但由于背景和历史原因，在桥梁及某些基础和承重墙的设计中也更常采用工作应力设计方法。

换算截面

根据钢筋混凝土梁的工作应力理论，钢筋和相邻混凝土的应变相等。因此，钢筋应力 f_s 是混凝土应力 f_c 的 n 倍，n 是钢的弹性模量 E_s 与混凝土弹性模量 E_c 的比值。钢上承受的力等于 nA_s。这表明在计算应力时，钢的面积可以等效为 n 倍混凝土的面积。

等效后的混凝土梁截面中，钢筋已被等效面积的混凝土所取代（图 5.2b）（在双钢筋梁和板中，应考虑混凝土的徐变和应力-应变图非线性的影响，采用系数为 $2n$ 对受压钢筋进行等效，但计算的应力应不超过允许的拉应力）。由于应力应变假定为随离中性轴的距离增加，传统的均质梁弹性理论适用于等效截面。截面特性，如中性轴的位置、转动惯量、截面模量 S 可按通常的方法计算，应力可由公式 $f=M/S$ 求得，其中 M 为弯矩。

混凝土梁的变形计算及准则

工作应力理论也可用于计算工作荷载下的变形，也就是说，弹性理论变形公式可用于钢筋混凝土梁。其中，有效转动惯量 I_e 按公式（5.1）计算

$$I_e = \left(\frac{M_{cr}}{M_a}\right)^3 I_g + \left[1 - \left(\frac{M_{cr}}{M_a}\right)^3\right] I_{cr} \leqslant I_g \tag{5.1}$$

式中　I_g——混凝土截面总惯性矩；

M_{cr}——开裂弯矩；

M_a——计算变形的弯矩；

I_{cr}——开裂混凝土（等效）截面。

若取 y_t 为总截面形心轴到受拉表面的距离，忽略钢筋，则混凝土的断裂弯矩按下式计算：

$$M_{cr} = \frac{f_r I_g}{y_t} \tag{5.2}$$

混凝土的断裂模数 $f_r = 7.5\sqrt{f'_c}$。公式（5.1）中，无论混凝土是否开裂，均考虑了混凝土截面惯性矩的变化。混凝土弹性模量 E_c 也可以计算。

这样计算得到的变形是假设荷载施加后立刻产生的变形。总变形如下

$$\Delta_{LT} = \Delta_L + \lambda_\infty \Delta_D + \lambda_t \Delta_{LS} \tag{5.3}$$

式中　Δ_L——活荷载产生的初始变形；

Δ_D——恒荷载产生的初始变形；

Δ_{LS}——持久荷载产生的初始变形；

λ_∞——恒荷载的乘数；

λ_t——持久荷载的乘数。

变形限制

ACI 规范建议对建筑物中的挠度进行以下限制：

不支撑且不附着可能被大挠度损坏的非结构元件的屋面板，活载荷下的最大立即挠度

不应超过 $L/180$，其中 L 是梁或板的跨度。

不支撑隔板且未连接到非结构元件的楼板，活载荷下的最大中间挠度不应超过 $L/360$。

如果楼板或屋面板支撑的构件可能会被大变形损坏，由活荷载产生的直接变形以及持久荷载的收缩徐变产生的变形不应超过 $L/480$。如果结构不太可能被大变形损坏，变形限值可以增加到 $L/240$。但构件应容许变形且采用足够的措施以避免结构构件变形造成的支撑破坏或非结构构件破坏。

Text B   Design of Prestressed Concrete

Prestressing is the application of permanent forces to a member or structure to counteract the effects of subsequent loading. Applied to concrete, prestressing takes the form of **precompression**, usually to eliminate disadvantages stemming from the weakness of concrete in tension.

Basic Principles of Prestressed Concrete

The usual prestressing procedure is to stretch high-strength steel and anchor it to the concrete, which resists the tendency of the stretched steel to shorten and thus is compressed. The amount of prestress used generally is sufficient to prevent cracking or sometimes to avoid tension entirely, under service loads. As a result, the whole concrete cross section is available to resist tension and bending, whereas in reinforced-concrete construction, concrete in tension is considered ineffective. Hence, it is particularly advantageous with prestressed concrete to use high-strength concrete.

Prestressed-concrete pipe and tanks are made by wrapping steel wire under high tension around concrete cylinders. Domes are prestressed by wrapping tensioned steel wire around the **ring girders**. Beams and slabs are prestressed linearly with **steel tendons** anchored at their ends or bonded to the concrete. Piles also are prestressed linearly, usually to counteract handling stresses.

Prestressed concrete may be either **pretensioned** or **posttensioned**. For pretensioned concrete, the steel is stretched before the concrete is placed around it and the forces are transferred to the concrete by bond. For posttensioned concrete, bars or tendons are sheathed in ducts within the concrete forms and are tensioned after the concrete attains sufficient strength.

The final precompression of the concrete is not equal to the initial tension applied to the tendons. There are both immediate and long-time losses, which should be deducted from the initial prestress to determine the effective prestress to be used in design. One reason high-tensioned tendons are used for prestressing is to maintain the sum of these losses at a small percentage of the applied prestress.

In determining stresses in prestressed members, the prestressing forces may be treated the same way as other external loads. If the prestress is large enough to prevent cracking under design loads, elastic theory may be applied to the entire concrete cross section.

For example, consider the simple beam in Fig. 5.3 (a). Prestress P is applied by a straight tendon at a distance e_1 below the neutral axis. The resulting prestress in the extreme surfaces throughout equals $P/A + Pe_1 c/I$, where P/A is average stress on a cross section and $Pe_1 c/I$, the bending stress (+represents compression, −represents tension), as indicated in Fig. 5.3 (c). If, now, stresses $+M_c/I$ due to downward-acting loads are superimposed at midspan, the net stresses in the extreme surfaces may become zero at the bottom and compressive at the top (Fig. 5.3c). Since the stresses due to loads at the beam ends are zero, however, the prestress is the final stress there. Hence, the top of the beam at the ends will be in tension.

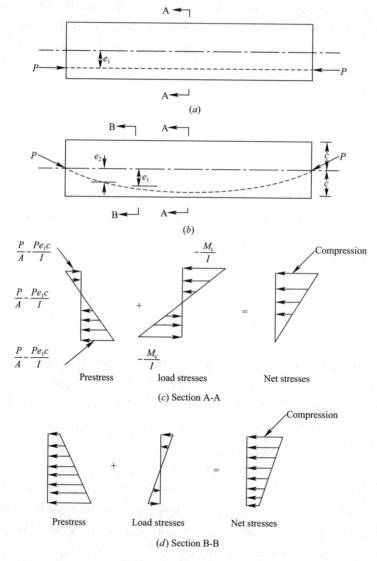

Fig. 5.3 Concrete beams

(a) Prestressed with straight tendons; (b) Prestressed with draped tendons; (c) Stress distribution at midspan; (d) Stress distribution for draped tendons at section between support and midspan. For straight tendons, net stress may be tensile near the supports

If this is objectionable, the tendons may be draped, or harped, in a vertical curve, as shown in Fig. 5.3 (b). Stresses at midspan will be substantially the same as before (assuming the horizontal component of P approximately equal to P), and the stress at the ends will be a compression, P/A, since P passes through the centroid of the section there. Between midspan and the ends, the cross sections also are in compression (Fig. 5.3d).

Losses in Prestress

The prestressing force acting on the concrete differs from the initial tension on the tendons by losses that occur immediately and over a long time.

Elastic Shortening of Concrete

In pretensioned members, when the tendons are released from **fixed abutments** and the steel stress is transferred to the concrete by bond, the concrete shortens because of the compressive stress. For axial prestress, the decrease in inches per inch of length may be taken as P_i/AE_c, where P_i is the initial prestress, kips; A is the concrete area, in^2; and E_c is the modulus of elasticity of the concrete, ksi. Hence, the decrease in unit stress in the tendons equals $P_i E_s/AE_c = nf_c$, where E_s is the modulus of elasticity of the steel, ksi; n the modular ratio; and f_c the stress in the concrete, ksi.

In posttensioned members, if tendons or cables are stretched individually, the stress loss in each due to compression of the concrete depends on the order of stretching. The loss will be greatest for the first tendon or cable stretched and least for the last one. The total loss may be approximated by assigning half the loss in the first cable to all. As an alternative, the tendons may be brought to the final prestress in steps.

Frictional Losses

In posttensioned members, there may be a loss of prestress where curved tendons **rub** against their enclosure. For harped tendons, the loss may be computed in terms of a curvature-friction coefficient μ. Losses due to unintentional misalignment may be calculated from a wobble-friction coefficient K (per lin ft). Since the coefficients vary considerably with duct material and construction methods, they should, if possible, be determined experimentally or obtained from the tendon manufacturer.

With K and m known or estimated, the friction loss can be computed from

$$P_s = P_x e^{Kl_x + \mu a} \qquad (5.4)$$

where P_s——force in tendon at prestressing jack, lb;

P_x——force in tendon at any point x, lb;

e——2.718;

l_x——length of tendon from jacking point to point x, ft;

a——total angular change of tendon profile from jacking end to point x, rad.

When $Kl_x + \mu a$ does not exceed 0.3, P_s may be obtained from

$$P_s = P_x(1 + Kl_x + \mu a) \qquad (5.5)$$

Slip at Anchorages

For posttensioned members, prestress loss may occur at the anchorages during the anchoring. For example, seating of wedges may permit some shortening of the tendons. If tests of a specific anchorage device indicate a shortening δl, the decrease in unit stress in the steel is $E_s dl/l$, where l is the length of the tendon.

Shrinkage of Concrete

Change in length of a member due to concrete shrinkage results over time in prestress loss. This should be determined from test or experience. Generally, the loss is greater for pretensioned members than for posttensioned members, which are prestressed after much of the shrinkage has occurred. Assuming a shrinkage of 0.0002 in/in for a pretensioned member, the loss in tension in the tendons will be

$$0.0002 E_s = 0.0002 \times 30000 = 6 \text{ksi} \tag{5.6}$$

Creep of Concrete

Change in length of concrete under sustained load induces a prestress loss over time. This loss may be several times the **elastic shortening**. An estimate of the loss may be made with a **creep coefficient** C_c, equal to the ratio of additional long-time deformation to initial elastic deformation, determined by test. Hence, for axial prestress, the loss in tension in the steel is $C_c n f_c$, where n is the modular ratio and f_c is the prestressing force divided by the concrete area. (Values ranging from 1.5 to 2.0 have been recommended for C_c.)

Relaxation of Steel

Decrease in stress under constant high strain occurs with some steels. For example, for steel tensioned to 60% of ultimate strength, relaxation loss may be 3%. This type of loss may be reduced by temporary overstressing, stabilizing the tendons by artificially accelerating relaxation and thus reducing the loss that will occur later at lower stresses.

Actual losses should be computed based on the actual initial stress level, type of steel (stressrelieved or low relaxation; wire, strand or bar), and prestressing method (pretensioned or posttensioned).

Design of Prestressed-Concrete Beams

This involves selection of shape and dimensions of the concrete portion, type and positioning of tendons, and amount of prestress. After a concrete shape and dimensions have been assumed, determine geometric properties: **cross-sectional** area, center of gravity, distances of extreme surfaces from the centroid, section moduli, and **dead load** of member per unit of length. Treat prestressing forces as a system of external forces acting on the concrete.

Compute bending stresses due to dead and live loads. From these, determine the magnitude and location of the prestressing force at points of maximum moment. This force must provide sufficient compression to **offset** the tensile stresses caused by the bending moments due to loads. But at the same time, it must not create any stresses exceeding the allowable values. Investigation of other sections will guide selection of tendons to be used and determine their position in the beam.

After establishing the **tendon profile**, prestressing forces, and **tendon areas**, check crit-

ical points along the beam under initial and final conditions, on removal from the forms, and during erection. Check ultimate strength in flexure and shear and the percentage of prestressing steel. Design anchorages, if required, and diagonal-tension steel. Finally, check **camber.**

The design may be based on the following assumptions. Strains vary linearly with depth. At cracked sections, the concrete cannot resist tension. Before cracking, stress is proportional to strain. The transformed area of bonded tendons may be included in pretensioned members and in posttensioned members after the tendons have been **grout**ed. Areas of open ducts should be deducted in calculations of section properties before bonding of tendons. The modulus of rupture should be determined from tests, or the cracking stress may be assumed as $7.5\sqrt{f_c'}$, where f_c' is the 28-day strength of the concrete, psi.

Prestressed beams should be checked by the strength theory. For bridge beams, the nominal strength should not be less than

$$\frac{U}{\varphi} = \frac{1.30}{\varphi}\Big[D + \frac{5}{3}(L+I)\Big] \tag{5.7}$$

where D——effect of dead load;

L——effect of design live load;

I——effect of impact;

φ——1.0 for factory-produced precast, prestressed members; 0.95 for posttensioned, cast-in-place members; 0.90 for shear.

The "Standard Specifications for Highway Bridges" (American Association of State Highway and Transportation Officials) recommend that prestressed-concrete flexural members be assumed to act as uncracked members subjected to combined axial and bending stresses under specified service loads. In pretensioned members and in posttensioned members after tendons have been grouted, the transformed area of bonded reinforcement may be taken into account in computations of section properties. For calculations of section properties before bonding of tendons, areas of open ducts should be deducted.

New Words and Expressions

[1] precompression ['priːkəm'preʃən] *n.* 预压;
[2] ring girders 环梁;
[3] steel tendons 钢筋;
[4] pretensioned [prɪ'tenʃnd] *adj.* 先张的;
[5] posttensioned [pəʊst'tenʃənd] *adj.* 后张的;
[6] losses in prestress 预应力损失;
[7] elastic shortening of concrete 混凝土弹性压缩;
[8] fixed abutments 固定支座;
[9] frictional losses 摩擦损失;
[10] rub [rʌb] *n. & v.* 摩擦;
[11] slip at anchorages 锚固滑移;

[12] shrinkage of concrete 混凝土的收缩；
[13] creep of concrete 混凝土的徐变；
[14] elastic shortening 弹性压缩；
[15] creep coefficient 蠕变系数；
[16] relaxation of steel 钢筋松弛；
[17] cross-sectional 横截面；
[18] dead load 静载荷，恒载，静重；
[19] offset ['ɒfset] n. 偏移量；
[20] tendon profile 预应力钢筋腱外形；
[21] tendon areas 肌腱区；
[22] critical points 临界点；
[23] camber ['kæmbə] n. 曲面。

LESSON 6 DESIGN OF STEEL STRUCTURE

Text A Introduction to Limit State Design

A Civil Engineering Designer has to ensure that the structures and facilities he designs are (i) fit for their purpose (ii) safe and (iii) economical and durable. Thus safety is one of the paramount responsibilities of the designer. However, it is difficult to assess at the design stage how safe a proposed design will actually be. There is, in fact, a great deal of uncertainty about the many factors, which influence both safety and economy. The uncertainties affecting the safety of a structure are due to

1. Uncertainty about loading;
2. Uncertainty about material strength;
3. Uncertainty about structural dimensions and behaviour.

These uncertainties together make it impossible for a designer to guarantee that a structure will be absolutely safe. All that the designer can ensure is that the risk of failure is extremely small, despite the uncertainties.

An illustration of the statistical meaning of safety is given in Fig. 6.1. Let us consider a structural component (say, a beam) designed to carry a given nominal load. Bending moments (B. M.) produced by loads are first computed. These are to be compared with the resistance or strength (R. M.) of the beam. But the resistance (R. M.) itself is not a fixed quantity, due to variations in material strengths that might occur between nominally same elements. The statistical distribution of these member strengths (or resistances) will be as sketched in (a).

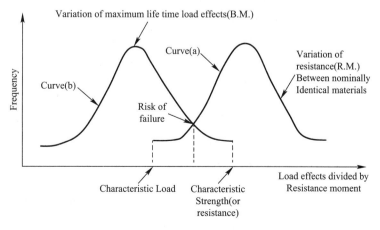

Fig. 6.1 Statistical meaning of safety

Similarly, the variation in the maximum loads and therefore load effects (such as bending moment) which different structural elements (all nominally the same) might encounter in their service life would have a distribution shown in (b). The uncertainty here is both due to variability of the loads applied to the structure, and also due to the variability of the load distribution through the structure. Thus, if a particularly weak structural component is subjected to a heavy load which exceeds the strength of the structural component, clearly failure could occur.

Unfortunately it is not practicable to define the probability distributions of loads and strengths, as it will involve hundreds of tests on samples of components. Normal design calculations are made using a single value for each load and for each material property and taking an appropriate safety factor in the design calculations. The single value used is termed as "Characteristic Strength or Resistance" and "Characteristic Load".

Characteristic resistance of a material (such as Concrete or Steel) is defined as that value of resistance below which not more than a prescribed percentage of test results may be expected to fall (For example the characteristic yield stress of steel is usually defined as that value of yield stress below which not more than 5% of the test values may be expected to fall). In other words, this strength is expected to be exceeded by 95% of the cases.

Similarly, the characteristic load is that value of the load, which has an accepted probability of not being exceeded during the life span of the structure. Characteristic load is therefore that load which will not be exceeded 95% of thetime.

Most structural designs are based on experience. If a similar design has been built successfully elsewhere, there is no reason why a designer may not consider it prudent to follow aspects of design that have proved successful, and adopt standardised design rules. As the consequences of bad design can be **catastrophic**, the society expects designers to explain their design decisions. It is therefore advantageous to use methods of design that have proved safe in the past. Standardised design methods can help in comparing alternative designs while minimising the risk of the cheapest design being less safe than the others. The regulations and guidelines to be followed in design are given in the Codes of Practices which help in ensuring the safety of structures.

The development of linear elastic theories in the 19th century enabled indeterminate structures to be analysed and the distribution of bending and shear stresses to be computed correctly. In the Working Stress Method (WSM) of design, the first attainment of yield stress of steel was generally taken to be the onset of failure as it represents the point from which the actual behaviour will deviate from the analysis results. Also, it was ensured that non-linearity and buckling effects were not present. It was ensured that the stresses caused by the working loads are less than an allowable stress obtained by dividing the yield stress by a factor of safety. The factor of safety represented a margin for uncertainties in strength and load. The value of factor of safety in most cases is taken to be around 1.67.

$$\text{Allowable Stress} = \frac{\text{Yield Stress}}{\text{Factor of Safety}} \qquad (6.1)$$

In general, each member in a structure is checked for a number of different combinations of loads. Some loads vary with time and this should be taken care of. It is unnecessarily severe to consider the effects of all loads acting simultaneously with their full design value, while maintaining the same factor of safety. Using the same factor of safety when loads act in combination would result in uneconomic designs. A typical example of a set of load combinations is given below, which accounts for the fact that the dead load, live load and wind load are all unlikely to act on the structure simultaneously at their maximum values:

(Stress due to dead load+live load)≤allowable stress

(Stress due to dead load+wind load)≤allowable stress

(Stress due to dead load+live load+wind)≤1.33 times allowable stress.

In practice there are severe limitations to this approach. The major limitation stems from the fact that yielding at any single point does not lead to failure. This means that the actual factor of safety is generally different from the assumed factor of safety and varies from structure to structure. There are also the consequences of material nonlinearity, nonlinear behaviour of elements in the post-buckled state and the ability of the steel components to tolerate high local stresses by yielding and redistributing the loads. The elastic theory does not consider the larger safety factor for statically indeterminate structures which exhibit redistribution of loads from one member to another before collapse. These are addresses in a more rational way in Limit State Design.

New Words and Expressions

catastrophic [ˌkætəˈstrɒfɪk] *adj.* 毁坏性的。

参考译文：极限状态设计简介

土木工程师需要确保所设计的结构、设施满足安全、适用、经济、耐久的要求。因此，安全是设计者的一项重要责任。然而，在设计阶段很难评估一个设计方案实际上会有多安全。实际上，结构存在诸多不确定性因素，这些因素既影响安全性又影响经济性。影响结构安全的不确定性因素包括：

1. 荷载不确定性；
2. 材料强度不确定性；
3. 结构尺寸及行为的不确定性。

这些不确定性综合作用使得难以确保结构绝对安全。设计者可以确定的是，尽管存在不确定性，但结构破坏风险特别小。

图6.1给出了安全的统计学意义。让我们考虑一个结构构件（比如梁）承受正态分布的荷载。首先可以计算得到荷载产生的弯矩。然后比较弯矩和梁抗力的大小。但由于同种构件的材料强度的离散性，梁的抗力并不是确定的值。曲线（a）画出了这些构件抗力的统计学分布情况。

类似的，名义上相同的不同结构构件在服役期能遇到的最大荷载及其产生的荷载效应（如弯矩）也服从曲线（b）所示的分布。这里的不确定性是由荷载的离散性和荷载分布的离散性产生的。因此，如果一个特别弱的构件遇到了特别重的荷载，超出了构件抗力，明显会出现破坏。

不幸的是通过成百上千次实验确定荷载和强度的概率分布不可行。通常的设计方法是对每一种荷载和每一种材料特性分别用一个定值来计算，再考虑一定的安全系数。这个定值称为"强度代表值或抗力代表值"和"荷载代表值"。

材料（如混凝土或钢材）的抗力代表值指的是具有一定保证率的值，该值与提前确定的百分比有关，实验中实际抗力低于抗力代表值的试件数量应低于该百分比（例如，钢材的屈服强度代表值是按5%确定的，实验中只有少于5%的试件的屈服强度低于该屈服强度代表值）。换句话说，95%以上的试件的强度会高于该抗力代表值。

类似的，荷载代表值是与概率相关的一个值，在结构整个生命周期中，荷载超越该代表值的概率低于一个可接受的百分比。荷载代表值实际按95%定义，结构寿命周期95%的时间中荷载不超过该代表值。

大部分的结构设计是基于经验的。如果别处已有类似结构的成功案例，那设计师可以参考该成功经验，并采用标准化的设计准则。由于不好的设计会导致灾难性的后果，公众期望设计者能解释他们的设计想法。因此，采用已被证明过安全性的设计方法是有利的。标准化设计方法可以帮助比较不同设计方案，避免最廉价的设计方案的安全性低于其他设计方案。设计规范也给出了需要遵从的条例和指南，可以帮助保障结构安全。

19世纪发展出来的线弹性理论提供了静不定结构的分析方法，并使得弯矩剪力的分布可以正确计算。在工作应力设计方法中，钢材的屈服点被确定为破坏准则，因为当超过这个点，结构的实际反应会和分析结果不同。非线性和失稳还没有给出。需要确保工作荷载产生的应力应低于屈服应力除以安全系数所得到的容许应力。安全系数代表对强度和抗力的不确定性留有多大余地。对于大部分情况安全系数取1.67。

$$容许应力 = \frac{屈服应力}{安全系数} \qquad (6.1)$$

总的来说，每一根构件承受的每一种荷载组合形式都需要验算。有些荷载会随时间变化，这也需要考虑在内。并不需要考虑所有荷载的最大值同时出现，不需要每种荷载采用同样的安全系数。考虑荷载组合时采用相同的安全系数会导致设计不经济。下文给出了部分典型的荷载组合，以说明结构的恒荷载、活荷载和风荷载的最大值同时发生的可能性很小。

（恒荷载+活荷载产生的应力）<容许应力

（恒荷载+风荷载产生的应力）<容许应力

（恒荷载+活荷载产+风荷载生的应力）<1.33倍的容许应力

在实际工程中这种设计方法存在几项缺陷。最主要的缺陷是某一点的屈服并不会导致破坏，这意味着实际的安全系数和假设的安全系数并不相同，而且因结构而异。由于材料非线性、构件在失稳后的非线性行为以及钢结构构件为了承受屈服后和应力重分布导致的高局部应力也会导致上述问题。弹性理论并不考虑静不定结构在倒塌前展现出来的荷载在构件间重新分布导致更大的安全系数问题。这些在极限状态设计中会用更合理的方式处理。

Text B Analysis Procedures and Design Philosophy

An improved design philosophy to make allowances for the shortcomings in the Working Stress Method was developed in the late 1970's and has been extensively incorporated in design standards and codes. The probability of operating conditions not reaching failure conditions forms the basis of Limit State Method (LSM). The Limit State is the condition in which a structure would be considered to have failed to fulfill the purpose for which it was built. In general two limit states are considered at the design stage and these are listed in Tab. 6.1.

Tab. 6.1 Types of limit states

Limit State of Strength	Limit State of Serviceability
Yielding, crushing and rupture	Deflection
Stability against buckling, overturning and sway	Vibration
Fracture due to fatigue	Fatigue checks (including reparable damage due to fatigue)
Brittle fracture	Corrosion

Limit State of Collapse is a catastrophic state, which requires a larger reliability in order to reduce the probability of its occurrence to a very low level. Limit State of Serviceability refers to the limit on acceptable service performance of the structure. Not all the limit states can be covered by structural calculations. For example, corrosion is covered by specifying forms of protection (like painting) and brittle fracture is covered by material specifications, which ensure that steel is sufficiently ductile.

The major innovation in the Limit State Method is the introduction of the partial safety factor format which essentially splits the factor of safety into two factors-one for the material and one for the load. In accordance with these concepts, the safety format used in Limit State Codes is based on probable maximum load and probable minimum strengths, so that a consistent level of safety is achieved.

Thus, the design requirements are expressedas follows:

$$F_d \leqslant S_d \tag{6.2}$$

Where F_d——value of internal forces and moments caused by the factored design loads;

$F_d = \gamma_f \times$ Characteristic Loads;

γ_f——partial safety factor for load (load factor);

S_d——factored design resistance as a function of the material design strength F_d;

$S_d = \gamma_m \times$ Characteristic strength;

γ_m——partial safety factor for material strength。

Both the partial safety factors for load and material are determined on a 'probabilistic basis' of the corresponding quantity. It should be noted that γ_f makes allowance for possible deviation of loads and also the reduced possibility of all loads acting together. On the other hand γ_m allows for uncertainties of element behaviour and possible strength reduction due to manufacturing tolerances and imperfections in the material. The partial safety factor for steel material failure by yielding or buckling γ_{m0} is given as 1.10 while for ultimate resistance it is given as $\gamma_{m1}=1.25$. For bolts and shop welds, the factor is 1.25 and for field welds it is 1.50.

Strength is not the only possible failure mode. Excessive deflection, excessive **vibration**, fracture etc. also contribute to Limit States. **Fatigue** is also an important design criterion for bridges, **crane girders** etc. Thus the following limit states may be identified for design purposes:

Collapse Limit States are related to the maximum design load capacity under extreme conditions. The partial load factors are chosen to reflect the probability of extreme conditions, when loads act alone orin combination. Stability shall be ensured for the structure as a whole and for each of its elements. It includes overall frame stability against overturning and sway, uplift or sliding under factored loads.

Serviceability Limit States are related to the criteria governing normal use. Unfactored loads are used to check the adequacy of the structure. These include **Limit State of Deflection, Limit State of Vibration, Limit State of Durability** and Limit **State of Fire Resistance**. Load factor, γ_f, of value equal to unity shall be used for all loads leading to serviceability limit states.

Fatigue Limit State is important where distress to the structure by repeated loading is a possibility. Stress changes due to **fluctuations** in wind loading normally need not be considered. When designing for fatigue, the load factor for action, γ_f, equal to unity shall be used for the load causing stress fluctuation and stress range.

Sections normally used in steel structures are I-sections, Channels or angles etc. which are called open sections, or rectangular or circular tubes which are called closed sections. These sections can be regarded as a combination of individual plate elements connected together to form the required shape. The strength of compression members made of such sections depends on their **slenderness ratio.** Higher strengths can be obtained by reducing the slenderness ratio i.e. by increasing the moment of inertia of the cross-section. Similarly, the strengths of beams can be increased, by increasing the moment of inertia of the cross-section. For a given cross-sectional area, higher moment of inertia can be obtained by making the sections thin-walled. However, the buckling of the plate elements of the cross section under compression/shear may take place before the overall column buckling or overall beam failure by lateral buckling or yielding. This phenomenon is called local buckling. Thus, local buckling imposes a limit to the extent to which sections can be made thin-walled.

Local buckling has the effect of reducing the load carrying capacity of columns and beams due to the reduction in **stiffness** and strength of the locally buckled plate elements. It is useful to classify sections based on their tendency to buckle locally before overall failure of the member takes place. The codes also specify the limiting width-thickness ratios $\beta=b/t$ for component plates, which enables the classification to be made. The cross-sections are classified into plastic, compact, semi-compact and slender depending upon their width-thickness ratios $\beta=b/t$ for component plates. This will be discussed in more detail in the chapter on beams.

Fabrication and erection are important aspects to be considered in the design of any steel structure. Fabrication includes the process of straightening, bending, cutting, machining and drilling. The difficult involved in performing these operation will have a major influence on the cost of the structure. Fabrication may be done either entirely in the stags, or entirely in the field or partly in both places. Similarly case of erection also influences the design.

It should be noted that the **code** gives only guidelines for design which when followed will reduce the probability of a structure collapsing. However, it is the designer's responsibility to ensure that the structure does not collapse due to loads or actions which are special to the particular structure, improper construction and erection techniques, mistakes in calculations etc.

New Words and Expressions
[1] vibration [vəˈbreɪʃ(ə)n] *n.* 振动；震动；颤动；
[2] fatigue [fətiːg] *n. & v.* 疲劳；
[3] crane girders 起重机梁；吊车梁；
[4] serviceability limit states 正常使用极限状态；
[5] limit state of deflection 挠度极限状态；
[6] limit state of vibration 振动极限状态；
[7] limit state of durability 耐久性极限状态；
[8] state of fire resistance 耐火状态；
[9] fluctuation [ˌflʌktjuˈeɪʃəns] *n.* 波动，涨落，起伏；
[10] slenderness ratio 长细比；
[11] stiffness [ˈstɪfnɪs] *n.* 刚度；硬度；
[12] code [kəʊd] *n.* 规范。

LESSON 7 FOUNDATION ENGINEERING

Text A Types of Failure in Soil at Ultimate Load

The lowest part of a structure that transmits its weight to the underlying soil or rock is the foundation. Foundations can be classified into two major categories—**shallow foundations** and **deep foundations.** Individual footings (Fig. 7.1), square or rectangular in plan, that support columns and strip footings that support walls and other similar structures are generally referred to as shallow foundations. **Mat foundations,** also considered shallow foundations, are reinforced concrete **slabs** of considerable structural **rigidity** that support a number of columns and wall loads. Several types of mat foundations are currently used. Some of the common types are shown schematically in Fig. 7.2 and include

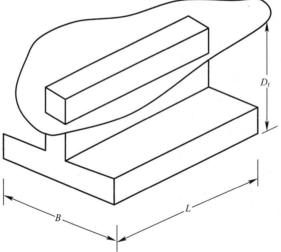

Fig. 7.1 Individual footing

1. Flat plate (Fig. 7.2a). The mat is of uniform thickness.
2. Flat plate thickened under columns (Fig. 7.2b).
3. Beams and slab (Fig. 7.2c). The beams run both ways, and the columns are located at the **intersections** of the beams.
4. Flat plates with **pedestals** (Fig. 7.2d).
5. Slabs with basement walls as a part of the mat (Fig. 7.2e). The walls act as stiffeners for the mat.

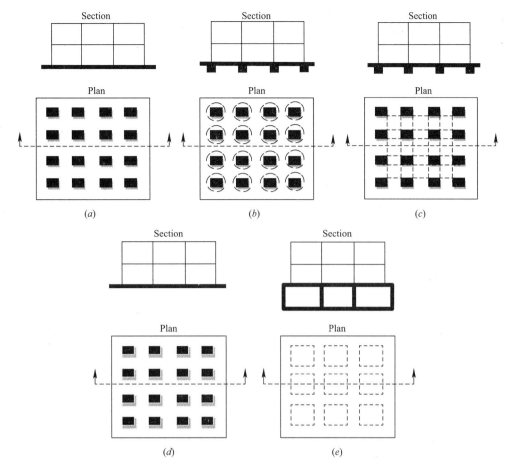

Fig. 7.2 Various types of mat foundations
(a) at plate; (b) at plate thickened under columns; (c) beams and slab;
(d) at plate with pedestals; (e) slabs with basement walls

When the soil located immediately below a given structure is weak, the load of the structure may be transmitted to a greater depth by **piles** and **drilled shafts**, which are considered **deep foundations**. This book is a compilation of the theoretical and experimental evaluations presently available in the literature as they relate to the load-bearing capacity and settlement of shallow foundations.

The shallow foundation shown in Fig. 7.1 has a width B and a length L. The depth of **embedment** below the ground surface is equal to D_f. Theoretically, when B/L is equal to zero (that is, $L=\infty$), a plane strain case will exist in the soil mass supporting the foundation. For most practical cases, when $B/L \leqslant 1/5$ to $1/6$, the plane strain theories will yield fairly good results. Terzaghi1 defined a shallow foundation as one in which the depth D_f is less than or equal to the width B ($D_f/B \leqslant 1$). However, research studies conducted since then have shown that D_f/B can be as large as 3 to 4 for shallow foundations.

Fig. 7.3 shows a shallow foundation of width B located at a depth of D_f below the ground surface and supported by dense sand (or stiff, clayey soil). If this foundation is subjected to a load Q that is gradually increased, the load per unit area, $q=Q/A$ (A=area of the foundation), will increase and the foundation will undergo increased settlement. When q becomes equal to q_u at foundation settlement $S=S_u$, the soil supporting the foundation undergoes sudden shear failure. The failure surface in the soil is shown in Fig. 7.3 (a), and the q versus S plot is shown in Fig. 7.3 (b). This type of failure is called a **general shear failure**, and q_u is the **ultimate bearing capacity**.

Note that, in this type of failure, a peak value of $q=q_u$ is clearly defined in the load-settlement curve.

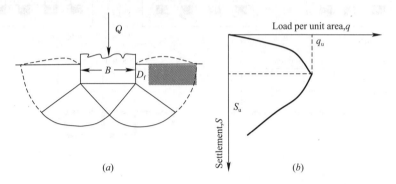

Fig. 7.3   General shear failure in soil

If the foundation shown in Fig. 7.3 (a) is supported by a medium dense sand or clayey soil of medium consistency (Fig. 7.4a), the plot of q versus S will be as shown in Fig. 7.4 (b).

Note that the magnitude of q increases with settlement up to $q=q'_u$, and this is usually referred to as the **first failure load**. At this time, the developed failure surface in the soil will be as shown by the solid lines in Fig. 7.4 (a). If the load on the foundation is further increased, the load-settlement curve becomes steeper and more erratic with the gradual outward and upward progress of the failure surface in the soil (shown by the jagged line in Fig. 7.4b) under the foundation. When q becomes equal to q_u (ultimate bearing capacity), the failure surface reaches the ground surface. Beyond that, the plot of q versus S takes almost a linear shape, and a peak load is never observed. This type of bearing capacity failure is called a **local shear failure**.

Shallow Foundations: Bearing Capacity and Settlement

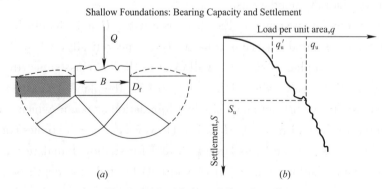

Fig. 7.4   Local shear failure in soil

Fig. 7.5 (a) shows the same foundation located on a loose sand or soft clayey soil. For this case, the load-settlement curve will be like that shown in Fig. 7.5 (b). A peak value of load per unit area q is never observed. The ultimate bearing capacity q_u is defined as the point where $\Delta S/\Delta q$ becomes the largest and remains almost constant thereafter. This type of failure in soil is called a **punching shear failure.** In this case the failure surface never extends up to the ground surface. In some cases of punching shear failure, it may be difficult to determine the ultimate load per unit area q_u from the q versus S plot shown in Fig. 7.5. DeBeer recommended a very consistent ultimate load criteria in which a plot of log $q/\gamma B$ versus log S/B is prepared (γ=unit weight of soil). The ultimate load is deened as the point of break in the log-log plot as shown in Fig. 7.6.

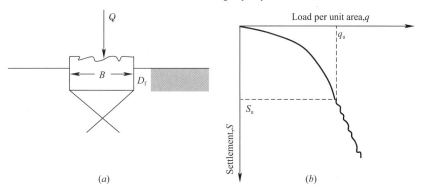

Fig. 7.5 Punching shear failure in soil

The nature of failure in soil at ultimate load is a function of several factors such as the strength and the relative compressibility of the soil, the depth of the foundation (D_f) in relation to the foundation width B, and the width-to-length ratio (B/L) of the foundation. This was clearly explained by Vesic, who conducted extensive laboratory model tests in sand. The summary of Vesic's findings is shown in a slightly different form in Fig. 7.7. In this figure D_r is the relative density of sand, and the hydraulic radius R of the foundation is defined as

$$R = \frac{A}{P} \tag{7.1}$$

Where, A=area of the foundation=BL; P=perimeter of the foundation=$2(B+L)$. Thus,

$$R = \frac{B}{2(B+L)} \tag{7.2}$$

for a square foundation $B=L$. So,

$$R = \frac{B}{4} \tag{7.3}$$

From Fig. 7.7 it can be seen that when $D_f/R \geqslant$ about 18, punching shear failure occurs in all cases irrespective of the relative density of compaction of sand.

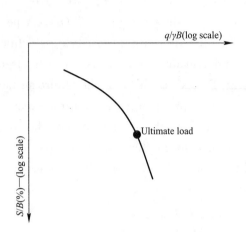
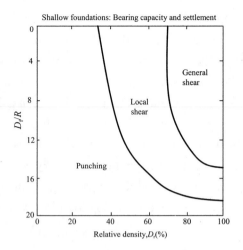

Fig. 7. 6 Nature of variation of $q/\gamma B$ with S/B in a log-log plot

Fig. 7. 7 Nature of failure in soil with relative density of sand D_r and D_f/R

Settlement at Ultimate Load

The settlement of the foundation at ultimate load S_u is quite variable and depends on several factors. A general sense can be derived from the laboratory model test results in sand for surface foundations ($D_f/B=0$) provided by Vesic and which are presented in Fig. 7. 8. From this figure it can be seen that, for any given foundation, a decrease in the relative density of sand results in an increase in the settlement at ultimate load. DeBeer

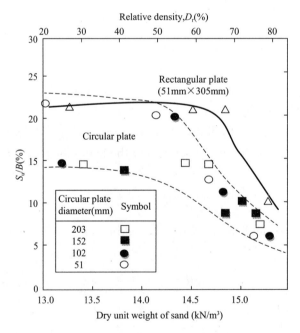

Fig. 7. 8 Variation of $\dfrac{S_u}{B}$ for surface foundation $\left(\dfrac{D_f}{B}=0\right)$ on sand

Source: From Vesic, A. S. 1973. Analysis of ultimate loads on shallow foundations.
J. Soil Mech. Found. Div. , ASCE, 99 (1): 45.

provided laboratory test results of circular surface foundations having diameters of 38mm, 90mm, and 150mm on sand at various relative densities (D_r) of compaction. The results of these tests are summarized in Fig. 7.9. It can be seen that, in general, for granular soils the settlement at ultimate load S_u increases with the increase in the width of the foundation B.

Based on laboratory and field test results, the approximate ranges of values of S_u in various types of soil are given in Table 7.1.

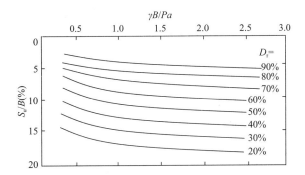

Fig. 7.9 DeBeer's laboratory test results on circular surface foundations on sand—variation of $\frac{S_u}{B}$ with $\frac{\gamma B}{Pa}$ and D_r

Note: B = diameter of circular foundation; Pa = atmospheric pressure $\approx 100 kN/m^2$; γ = unit weight of sand.

Shallow foundations: Bearing capacity and settlement

Tab. 7.1 **Approximate ranges of Su**

Soil	$\frac{D_f}{B}$	$\frac{S_u}{B}(\%)$
Sand	0	5~12
Sand	Large	25~28
Clay	0	4~8
Clay	Large	15~20

For a given foundation to perform to its optimum capacity, one must ensure that the load per unit area of the foundation does not exceed a limiting value, thereby causing shear failure in soil. This limiting value is the ultimate bearing capacity q_u. Considering the ultimate bearing capacity and the uncertainties involved in evaluating the shear strength parameters of the soil, the allowable bearing capacity q_{all} can be obtained as

$$q_{all} = \frac{q_u}{FS} \quad (7.4)$$

A factor of safety of three to four is generally used. However, based on limiting settlement conditions, there are other factors that must be taken into account in **deriving** the allowable bearing capacity. The total settlement S_t of a foundation will be the sum of the following:

1. Elastic, or immediate, settlement S_e (described in section 7.3).

2. Primary and secondary consolidation settlement S_c of a clay layer (located below the groundwater level) if located at a reasonably small depth below the foundation.

Most building codes provide an allowable settlement limit for a foundation, which may be well below the settlement derived corresponding to q_{all} given by Eq. (7.4). Thus, the bearing capacity corresponding to the allowable settlement must also be taken into consideration.

A given structure with several shallow foundations may undergo uniform settlement (Fig. 7.10a). This occurs when a structure is built over a very rigid structural mat. However, depending on the loads on various foundation components, a structure may experience **differential settlement**. A foundation may undergo uniform **tilt** (Fig. 7.10b) or nonuniform settlement (Fig. 7.10c). In these cases, the angular distortion Δ can be defined as

$$\Delta = \frac{S_{t(max)} - S_{t(min)}}{L'} \quad \text{(for uniform tilt)} \tag{7.5}$$

and

$$\Delta = \frac{S_{t(max)} - S_{t(min)}}{L'} \quad \text{(for nonuniform tilt)} \tag{7.6}$$

Limits for allowable differential settlements of various structures are also available in building codes. Thus, the final decision on the allowable bearing capacity of a foundation will depend on Fig. 7.10 (a) the ultimate bearing capacity, (b) the allowable settlement, and (c) the allowable differential settlement for the structure.

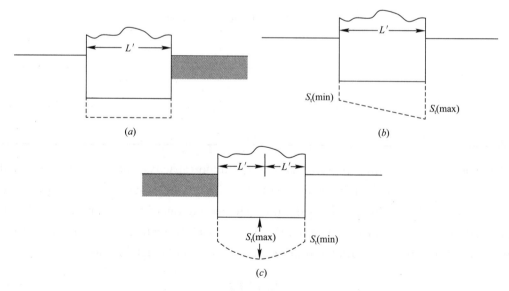

Fig. 7.10 The allowable bearing capacity of a foundation
(a) Uniform settlement; (b) Uniform tilt; (c) Nonuniform settlement

New Words and Expressions
[1] shallow foundations 浅基础;

［2］ deep foundations 深基础；

［3］ mat foundations 筏式基础；

［4］ slabs [slæbz] n. 砖；厚板；

［5］ rigidity [rɪˈdʒɪdətɪ] n. 刚性；刚度；

［6］ intersection [ɪntəˈsekʃnz] n. 交叉口；

［7］ pedestal [ˈpedɪstəlz] n. 基座；底座；

［8］ pile [paɪlz] n. 桩；

［9］ drilled shafts 钻孔桩井；

［10］ embedment [ɪmˈbedmənt] n. & v. 埋置；埋入；预埋件；

［11］ general shear failure 整体剪切破坏；

［12］ ultimate bearing capacity 极限承载力；

［13］ first failure load 初始破坏载荷；

［14］ local shear failure 局部剪切破坏；

［15］ punching shear failure 冲剪破坏；

［16］ tilt [tɪlt] v. & n. 倾斜；

［17］ differential settlement 不均匀沉降。

参考译文：极限荷载作用下土体破坏模式

基础位于结构最下部，将结构的重量传递到下卧的土体或岩石。基础可以分为浅基础和深基础两大类别。独立基础（图 7.1）通常为平面正方形或矩形，用于支撑柱子，条形基础用于支撑墙或类似结构，这些通常称为浅基础。筏形基础也是一种浅基础，它由具有一定刚度的钢筋混凝土板组成，用于支撑若干根柱子或若干片墙体。有几种筏形基础现在正在使用。图 7.2 显示了几种常用的筏形基础，它们包括：

1. 平板式筏形基础（图 7.2a），板厚均匀。
2. 柱下加厚筏形基础（图 7.2b）。
3. 梁板式筏形基础（图 7.2c），梁双向交叉布置，柱位于梁的交叉点上。
4. 带有底座的筏板基础（图 7.2d）。
5. 将地下室侧墙作为加厚版的筏形基础（图 7.2e），墙体起到增加基础刚度的作用

如果结构下方浅层土体强度较低，结构荷载可以通过桩或钻孔桩井传递到更深处，这称为深基础。本书整理了已有文献中关于浅基础承载力和沉降相关的理论与实验的评估方法。

图 7.1 所示的浅基础宽度为 B，长度为 L，埋置深度为 D_f。理论上当 B/L 等于 0 时（即 L 为无穷大），土体对于基础的支撑作用属于平面应变问题。对于大部分实际情况，当 B/L 介于 1/5～1/6 时，根据平面应变理论可以得到合理的结果。太沙基将浅基础定义为 D_f 小于等于宽度 B（$D_f/B \leqslant 1$）。然而，后续研究表明 D_f/B 达到 3～4 的基础也可以视作为浅基础。

图 7.3 所示宽度为 B 的浅基础位于埋置深度为 D_f 的位置，持力层为密实砂（或硬黏土）。如果这个基础受到的荷载 Q 逐渐增加，单位面积承受的荷载 $q=Q/A$（A 为基础面积）以及基础沉降也会逐渐增加。当 $q=q_u$ 时，$S=S_u$，支撑基础的土体会产生突然的剪切

破坏。图7.3（a）显示了土体破坏的表面，图7.3（b）给出了q与S关系曲线。这种破坏称为整体剪切破坏，q_u即为极限承载力。

需要注意的是，在这种破坏模式下，可在荷载沉降曲线上很明显地确定q的峰值q_u。

如果图7.3（a）所示的基础是由中密砂或黏性土（图7.4a）支撑的，q-S曲线形状将如图7.4（b）所示。

注意q的幅值随沉降增加直至$q=q'_u$，这通常称为初始破坏荷载。此时，土体破坏面如图7.4（a）中的实线所示。如果基础上的荷载进一步增加，荷载-沉降曲线将随着基础下面土体破坏面逐渐朝外朝上的变化而变得更陡、更不规则（如图7.4b中锯齿线所示）。当q等于q_u（极限承载力）时，破坏面到达地表面。之后，q-S曲线几乎为直线，没有峰值荷载。这种破坏模式称为局部剪切破坏。

图7.5（a）表示同一基础位于松砂或软黏土上的情况。在这种工况下，荷载-沉降曲线如图7.5（b）所示。无法得到单位面积荷载q的峰值。极限承载力q_u定义为$\Delta S/\Delta q$达到最大值且之后基本保持定值的点。土的这种破坏模式称为冲剪破坏。在这种情况下，破坏面不会延伸到地表。在一些冲剪破坏中，从图7.5所示的q-S曲线中确定极限承载力q_u比较困难。DeBeer建议采用统一极限荷载准则，该方法需要画出$\log q/\gamma B$和$\log S/B$（γ表示土的重度）关系曲线。极限荷载定义为图7.6所示的log-log图的断点。

土在极限荷载下的破坏规律可用考虑多种因素影响的函数表示，例如土的强度与压缩性、基础埋深（D_f）与宽度B的关系，以及基础的长宽比（B/L）。Veisc在实验室中对砂土进行了大量实验并对此作了清晰的解释。图7.7中用略有不同的形式总结了Veisc的研究成果。此图中，D_r是砂土密实度，基础的水力半径R定义为：

$$R = \frac{A}{P} \tag{7.1}$$

式中　　A——基础面积，$A=BL$；

　　　　P——基础周长，$P=2(B+L)$。

因此，

$$R = \frac{B}{2(B+L)} \tag{7.2}$$

对于方形基础$B=L$，则：

$$R = \frac{B}{4} \tag{7.3}$$

从图7.7可以看到，当D_f/R约大于等于18时，不论砂土密实度如何均会发生冲剪破坏。

极限荷载下的基础沉降

基础在极限荷载作用下产生的沉降S_u受很多因素影响，波动范围较大。Vesic将基础置于砂土表面（$D_f/B=0$）进行室内模型试验，图7.8所示的试验结果可以反映一定的规律。如图所示，对于任一给定基础，砂土密实度的降低导致极限荷载作用下的沉降增加。DeBeer提供了在密实度（D_r）砂土上的圆形基础的测试结果，基础的直径分别为38mm、90mm和150mm。图7.9总结了这些结果。可以看出，对于粒状土体，极限荷载作用下的沉降S_u随着基础宽度B增加而增加。

基于实验和现场测试得到的结果，表7.1给出了不同土的S_u的大致范围。

为使基础能够发挥其最优承载力,需要保证基础单位面积承受的荷载不超过限值,以免产生土体的剪切破坏。这个限值就是极限承载力 q_u。考虑到确定极限承载力及土的剪切强度存在不确定性,容许承载力 q_{all} 可以通过下式得到:

$$q_{all} = \frac{q_u}{FS} \tag{7.4}$$

通常安全系数取值在 3~4 之间。然而,根据基础沉降限制条件,确定容许承载力时必须考虑其他因素。基础总沉降 S_t 是下列沉降之和:

1. 弹性或瞬时沉降 S_e(如 7.3 节所述)。
2. 如果基础下一定深度内存在软弱下卧层,则黏土(地下水位线以下的)主固结与次固结产生的沉降 S_c 也需计算在内。

大部分建筑规范对基础的容许沉降限值给出了规定,规定限值可能远低于公式 (7.4) 得到的 q_{all}。因此,还需要考虑容许沉降所对应的承载力。

一个结构如果由几个浅基础支撑时可能产生均匀沉降(图 7.10a)。当结构建在刚性地基上时通常是这种情况。但是,当各基础承担荷载不同时,结构也可能产生不均匀沉降。基础可能产生均匀倾斜(图 7.10b)或者不均匀沉降(图 7.10c)。在这些情况下,转角位移 Δ 可以定义为

$$\Delta = \frac{S_{t(max)} - S_{t(min)}}{L'} \quad \text{(for uniform tilt)} \tag{7.5}$$

和

$$\Delta = \frac{S_{t(max)} - S_{t(min)}}{L'} \quad \text{(for nonuniform tilt)} \tag{7.6}$$

建筑规范给出了不同结构的容许的不均匀沉降限值。因此,基础的最终的容许承载力由图 7.10 (a) 所示的极限承载力、(b) 所示的容许沉降、(c) 所示的容许不均匀沉降确定。

Text B Piled Foundation Choice

The first decision in considering a foundation design is whether piles are required or not. In some cases there may be alternative solutions, for which the costs may be compared with those of a piled foundation. In other cases, the bearing capacity of the soil at the foundation level may be satisfactory but, owing to high loadings, piles are required to keep settlement within acceptable limits. It is important to be clear about the reasons for using bearing piles before weighing the relative merits of using steel or concrete types of driven pile, because there are some essential differences in behaviour that may favour one or the other pile type for a particular project.

Bearing piles are used mostly for supporting vertical loads and for this purpose the main requirements are to

- Restrict average settlement to a low value.
- Minimise differential settlement.
- Achieve an adequate factor of safety or load factor against foundation failure.

Many technical and cost-benefit factors affect the selection of the most appropriate type of pile for a given structure. Very broadly, these factors can be divided into those related to

- Site location and operating conditions.
- Type of soil and ground water level during installation.
- Type and size of the loads to be supported by the foundation.
- Type of structure, e.g. land or marine.
- Effect of the pile type on overall construction programme and cost.

In some circumstances there will be additional technical factors that affect the choice of pile, for instance when overturning moments due to wind forces on a tall building have to be resisted, or when severe scouring of a river bed may expose piles supporting a bridge pier.

Where piles have to resist tensile loading or absorb energy in bending, as in marine dolphins for ship impact, and in integral bridge piers for vehicle impact, there are special requirements to be considered which favour the selection of steel piles. In particular, the ductility of steel piles creates an elastic 'compliance' with the superstructure to absorb the impact energy by deflection.

The cost-benefit factors which may favour the choice of steel piles include

- **Total cost of the foundation**, where it is important that the comparison between pile types is related to the total construction cost including installation and not just the cost of the pile material.
- **Total construction time**, where use of driven steel piling can result in a shorter construction period and an earlier project completion date.
- **Environmental constraints**, where the noise and **vibration** caused during steel pile driving has now been reduced by developing new installation equipment to be within limits stated in UK legislation.
- **Sustainability issues**, where steel bearing piles are easy to extract from the ground at the end of structure life and can be reused or recycled so reducing the whole life cost of the building.

Many of the above factors are interrelated, and all require consideration in arriving at the most suitable pile type for a given situation. Broad guidance only is possible in this publication, as each project requires individual examination. For specific technical advice or product information, the organisations listed in Appendix A of this publication should be contacted.

There is no single pile type that is both technically and economically appropriate for every structure, site or set of soil conditions. Owing to the many different types of project and construction situations, there will always be a need for a variety of pile types, so selection is an exercise of judgement.

Knowledge about the installation and in-service performance of steel bearing piles has progressed over the last 40 years due to increased usage worldwide, particularly in the

USA, Japan and in European countries particularly Norway, Finland, Holland, Belgium and Denmark. Research work for the offshore industry has been carried out and reported in the UK and the transfer of this knowledge was considered beneficial for UK onshore application.

The trend towards increased foundation loads is well catered for by steel bearing piles. H-piles are capable of carrying loads of up to 4,400 kN. Steel piles offer many advantages compared to other types including

- Reduced foundation construction time and site occupation.
- Reliable section properties without need for onsite pile integrity checking.
- Ductility also gives high resistance to lateral loads for marine structures and compliance in integral bridge foundations.
- Larger wall surface area giving better friction capacity than equivalent diameter concrete pile
- Higher end bearing resistance in **granular** soils and rocks **mobilised** by pile driving as compared to boring.
- Closer spacing possible and therefore smaller pile caps.
- Pile load capacity can be confirmed during driving by Dynamic Pile Analysis (DPA) on every pile driven.
- Low displacement of adjacent soil during driving.
- No arisings and therefore no spoil disposal offsite
- Easily extracted at end of working life.
- Reusable or recyclable following extraction to meet Government objectives in sustainable construction

Steel piles have clear-cut advantages on projects such as on river or estuary crossings where soils are typically granular and **waterlogged** and unsuitable for satisfactory pile boring, or where soft recent low bearing strength alluvium overlies bedrock. On cohesive soil sites, there is a wide selection of acceptable pile types and other construction aspects will govern.

Nowadays, steel piling is an attractive and competitive alternative for permanent foundations owing to the research and development in piling technology and changes in the construction industry supply chain. These can be described under three broad headings, **durability**, performance and economy.

The subject of **corrosion** and steel protection has received **substantial** attention both in the UK and abroad over the last 40 years. There is now adequate knowledge on corrosion rates, **coatings** selection and specifications to permit the designer to make a reasoned judgement on the provision for corrosion prevention. Such information is readily available from Corus publications. In addition, the corrosion guidance sections of BS 8002, BS 6349, Eurocode 3: Part 5 (EN 1993-5) and in document BD 42 (part of the **Design manual for roads and bridges**) have embodied earlier research, and further revisions are in progress.

Reliable load capacity and driveability predictions are essential for the confident design and installation of driven piling. These topics have been poorly covered in most foundation and piling design textbooks and this publication therefore provides practical advice for the guidance of designers.

It was deemed appropriate to examine piling technology used in the offshore construction sector, where there is a body of research and accepted practice, and to transfer relevant practices to the onshore sector. The offshore design methods are simple in concept and the principles involved can be readily understood. They have been used with success in minimising foundation installation costs and the steel **tubular** piles have performed well for decades on offshore fixed structures.

For economic pile design, the methods require knowledgeable judgement of soil parameters and this, in turn, requires high quality soils data. Such data is obtainable using routine site investigation techniques, but care must be exercised in the soil sampling and testing specifications, in order to ensure that data collected on soil properties is relevant to driven steel piles as well as to bored concrete piles. In particular, there must be more emphasis on **in situ penetration** testing.

The differential in cost between concrete and steel construction has decreased steadily over recent years; the costs of site labour and concreting materials have increased, whereas the cost of steel has decreased in real terms. In addition, with the advent of 'Design and Build' contracts for civil engineering work, there is more incentive for innovative design to permit cheaper overall construction by incorporating the piled foundation into the structure concept rather than leaving it separate.

Constructing in steel permits **prefabrication** of larger, but still easily erectable, high quality structural elements that can save construction time; this is an increasingly attractive project consideration.

In foundations and basements, steel bearing piles are compatible and easily connectable to the steel frame of a building thereby permitting savings in overall construction costs. Progress has also been made in more effective connection between reinforced concrete superstructures and steel piling using welded-on shear **studs** or angles, **hoop** bar connectors and careful detailing in composite connections in bridge engineering. For steel intensive basement construction, cost savings of up to 40% have been reported by designers.

Steel foundation piles are **ductile** and can **deflect** to absorb energy in marine applications producing a saving in structural section.

New Words and Expressions
[1] total cost of the foundation 基础总成本；
[2] total construction time 施工总工期；
[3] environmental constraints 环境制约；
[4] vibration [vaɪˈbreɪʃn] v. & n. 振动；震动；颤动；

[5] sustainability issues 可持续发展问题；
[6] in situ penetration 原地渗透；
[7] prefabrication [ˌpriːfæbrɪˈkeɪʃn] *n.* 预制件；
[8] hoop [huːp] *n.* 卡箍；环；
[9] ductile [ˈdʌktaɪl] *adj.* 延性的；可延展的；易延展的；
[10] deflect [dɪˈflekt] *v.* 使转向；使偏向；使偏斜。

Text C　Sustainability and Environment

The world's available supply of construction materials is becoming more scarce and consequently more difficult and expensive to source and supply. Western governments have agreed to encourage more recycling of construction material in order to reduce the impact of mining more **ore** and aggregate, and to reduce the volume of waste construction materials from **demolition** of old buildings.

Steel is the world's most recycled material and is 100% recyclable. In 2003, 965 million tonnes of steel were produced worldwide and approximately 43% of that was from recycled **scrap** steel.

The use of scrap is also essential to the efficient production of the stronger higher grade steels and it therefore has a commercial value that makes recycling economically viable. The supply chain for scrap is well established (see the SCI publication **Environmental assessment of steel piling**).

When assessing the environmental impact of construction, it is important to consider the practicality and cost of removal of the structure at the end of its useful life and the disposal of the demolished materials. The construction industry, in common with many other industries, is now being encouraged to develop new processes that will allow more materials to be recycled or reused, helping to conserve natural resources and reduce waste.

Steel piling benefits from being easy to extract from the ground during demolition of previous structures, or after its temporary use as part of the construction process. Extraction equipment includes vibration hammers working under a pullout force from **cranes** and special high load jacking frames that can pull out the longer bearing piles. This facility creates an additional environmental benefit from being able to easily restore a previous building site to a 'greenfield' state without any remaining contamination below ground. The steel piles can either be reused or recycled.

Concrete piles on the other hand are difficult to demolish or extract and the process is therefore time consuming and expensive. On many sites the degree of **contamination** with concrete piles is so expensive to remove that developers have been deterred from using that 'brownfield' site and have used a 'greenfield' site instead. On some 'brownfield' sites, the new piled foundation has been interwoven through the old concrete piles, creating more contamination and rendering the site much worse for any future redevelopment. The large diameter bored concrete under-reamed piles that have often been used on inner city sites such as in London and Manchester, are particularly difficult to remove.

New Words and Expressions
[1] ore [ɔː(r)] n. 矿石；
[2] demolition [ˌdeməˈlɪʃn] n. 拆除；拆毁；
[3] scrap [skræp] n. 废料；
[4] environmental assessment of steel piling 钢桩环境评估；
[5] crane [kreinz] n. 塔吊；
[6] contamination [kənˌtæmɪˈneɪʃən] n. 污染。

LESSON 8　TUNNEL ENGINEERING

Text A　Tunnel Engineering

Tunnel engineering makes possible many vital underwater and underground facilities. Unique design and construction techniques are involved because of the necessity of protecting the constructors and users of these facilities from alien environments. These facilities must be built to exclude the materials through which they pass, including water. Often, they have to withstand high pressures. And when used for transportation or human occupancy, tunnels must provide adequate lighting and a safe atmosphere, with means for removing pollutants.

Tunnels are constructed using many methods, depending upon the kind of soil and/or rock through which they will pass, their size, how deep they need to be, and the obstructions that may be encountered along the route. These methods include **cut-and-cover construction, drill and blast, tunnel boring machine** (TBM), **immersion of prefabricated tunnels**, and **sequential excavation methods** (SEM). More specialized methods, such as **ground freezing** and **tunnel jacking**, are used less frequently and often under very difficult conditions. Compressed air working has become uneconomical because of working hour restrictions, time for decompression that results from high working pressures (over 40 psi is not unusual), union labor agreements for work under compressed air, and high workmen's compensation and health benefit rates. Occasional entry under compressed air may still be required, such as to clear obstructions ahead of a tunnel boring machine, or to perform essential maintenance on parts of such a machine.

The design approach to underground and underwater structures differs from that of most other structures. Internal space, design life, and other requirements for the tunnel must first be defined. Geological and environmental data must then be collected. **Critical design loading conditions** must then be established, including acceptable conditions of the tunnel following extreme events (for example, how long before the tunnel is reusable). **Appropriate construction methods** are then evaluated to determine the most appropriate to meet the established criteria, conditions, and cost. The methods under consideration should include both temporary and permanent excavation support systems as well as the structures itself. Design standards and codes of practice apply primarily to above-ground structures, so that care should be used in their application to underground and underwater structures.

Clearances for Tunnels

Clearance in a tunnel is the least distance between the inner surfaces of the tunnel necessary to provide space between the closest approach of vehicles or their cargo or **pedestrian traffic** and those surfaces. Minimum tunnel dimensions are determined by adding the minimum clearances established for a tunnel to the dimensions selected for the type of traffic to be accommodated in the tunnel and the space needed for other requirements, such as **ventilation** ducts and **pipelines**.

Clearances for Railroad Tunnels. Individual railroads have different standards to suit their equipment.

In rail tunnels, clearances for personnel are required on both sides where niches are not provided. These clearances should be at least 6ft 8in or 2m high and 30in wide each side of the vehicle clearance diagram, although a 24in minimum is permitted on some lines. In highway tunnels, a 3ft or 0.9m clearance from face of curb is used where walkways are provided. In both road and rail tunnels, it is common practice to provide a walkway along the common wall between adjacent ducts to facilitate emergency evacuation between ducts and to prevent people from emerging directly into the path of oncoming traffic.

On curved tracks, the clearances should be increased to allow for overhang and tilting of an 85-ft-long car, 60ft c to c of trucks, and a height of 15ft 1in above top of rail. (Distance from top of rails to top of ties should be taken as 8in.)

Clearances for Rapid-Transit Tunnels. There are no general standards for clearances in **rapid-transit tunnels**. Requirements vary with size of rolling stock used in the system.

Clearances for Highway Tunnels. The American Association of State Highway and Transportation officials (AASHTO) has established standard horizontal and vertical clearances for various classes of highways. These have been modified and expanded for the Interstate Highway System under the jurisdiction of the Federal Highway Authority (FHWA).

For rural and most urban parts of the Interstate Highway System, a 16-ft vertical clearance is required.

Since construction costs of tunnels are high, clearance requirements are usually somewhat reduced. Although some older 2-lane tunnels have used roadway widths of 21ft between curbs for unidirectional traffic and 23ft for bi-directional traffic, usually with speed restrictions, these widths no longer meet current standards for 12ft or 3.6m lanes. Full width shoulders are rarely provided due to cost, but at least an additional 1ft is provided adjacent to each curb. Wider shoulders or sight shelves may be required around horizontal curves to comply with sight distance requirements. A minimum distance between walls of 30ft is a common requirement. Resurfacing within tunnels is rarely permitted without first removing the old surfacing, so no allowance for resurfacing is required for overhead clearance. It is usual in tunnels to provide overhead lane signals to show which lanes are open to traffic in the direction of travel, so extra overhead allowance is required for these, and when appropriate also for lighting, overhead signs, jet fans for ventilation, and any other **ceiling-mounted** items. Minimum overhead traffic clearances depend upon which alternative routes are available for over-height vehicles and the classification of the highway, but accepted values usually lie between 14ft and 5.1m. Additional height may be required on vertical curves to allow for long trucks. Additional space may be required for ventilation, ventilation equipment, and ventilation ducts.

Alignment and Grades for Tunnels

Alignment of a tunnel, both horizontal and vertical, generally consists of straight lines connected by curves. Minimum grades are established to ensure adequate drainage. Maximum grades depend on the purpose of the tunnel. Construction of a tunnel in the upgrade direction is preferred whenever possible, since this permits water to drain away from the face under construction.

Alignment and Grades for Railroad Tunnels. Straight alignments and grades as low as possible, yet providing good drainage, are desirable for train operation. But overall construction costs must be taken into account. Grades in curved tunnels should be compensated for **curvature**, as is done for **open lines**. In general, maximum grades in tunnels should not exceed about 75% of the ruling grade of the line. This grade should be extended about 3000ft below and 1000ft above the tunnel.

Short (under 2500ft), unventilated tunnels should have a constant grade throughout. Long, ventilated tunnels may require a high point near the center for better drainage during construction if work starts from two headings.

Radii of curves and superelevation of tracks are governed by maximum train speeds.

Alignment and Grades for Rapid-Transit Tunnels. Radii of curvature and limiting grades are governed by operating requirements. The New York City IND Subway has a 350-ft minimum radius, with transition curves for radii below 2300ft. Maximum grades for this system are 3% between stations and 1.5% for turnouts and crossovers. The San Francisco BART system is designed for train speeds of 80mi/h. Relation of speed to radius and superelevation of track for horizontal curves is determined by

$$E = \frac{4.65V^2}{R} - U \tag{8.1}$$

Where E——superelevation, in;
R——radius, ft;
V——train speed, mi/h;
U——unbalanced superelevation, which should not exceed $2^{3/4}$ in optimum or 4 in as an absolute maximum.

For 80mi/h design speed, the radius with an optimum superelevation would be 5000ft. For a maximum permissible superelevation of $8^{1/4}$ in, a minimum radius of 3600ft would be required. The absolute minimum radius for yards and turnouts is 500ft. Maximum line grade is 3.0% and 1.0% in stations. To ensure good drainage, grade should preferably be not less than 0.50%.

Alignment and Grades for Highway Tunnels. For tunnels under **navigable** water carrying heavy traffic, upgrades are generally limited to 3.5%; downgrades of 4% are acceptable. For lighter traffic volumes, grades up to 5% have been used for economy's sake. Between governing navigation clearances, grades are reduced to a minimum adequate for drainage, preferably not less than 0.25% longitudinally and a **cross slope** of 1.0%. For long rock tunnels with two-way traffic, a maximum grade of 3% is desirable to maintain reasonable truck speeds. Additional climbing lanes for slower traffic may be required when grades exceed 4%.

Radii of curvature should match tunnel design speeds. Short radii require superelevation and some widening of roadway to provide for overhang and sight distance.

Preliminary Investigations

Surveys should be made to establish all **topographical features** and locate all surface and **subsurface structures** that may be affected by the tunnel construction. For underwater tunnels, soundings should be made to plot the bed levels.

Knowledge of geological conditions is essential for all tunnel construction but is of primary importance for rock tunnels. Explorations by borings and **seismic reflection** for soft ground and underwater tunnels are readily made to the extent necessary. For rock tunnels, particularly long ones, however, possibilities for borings are often limited. A thorough investigation should be made by a geologist familiar with the area. This study should be based on a careful surface investigation and examination of all available records, including records of other construction in the vicinity, such as previous tunnels, mines, quarries, open cuts, shafts, and borings. The geologist should prepare a comprehensive report for the guidance of designers and contractors.

For soft ground and underwater tunnels, borings should be made at regular intervals. They should be spaced 500 to 1000ft apart, depending on local conditions. Closer spacing should be used in areas of special construction, such as ventilation buildings, **portals**, and cut-and-cover sections. Spoon samples should be taken for soil classification, and undisturbed samples, where possible, for laboratory testing. Samples not needed in the laboratory, boring logs, and laboratory reports should be preserved for inspection by contractors. Density, shear and compressive strength, and plasticity of soils are of special interest.

All borings should be carried below tunnel invert. For pressure face tunnels, borings should be located outside the tunnel **cross section.**

For rock tunnels, as many borings as practicable should be made. Holes may be inclined, to cut as many layers as possible. Holes should be carried below the invert and may be staggered on either side of the center line, but preferably outside the tunnel cross section to prevent annoying water leaks. Where formations striking across the tunnel have **steep dips, horizontal borings** may give more information; borings 2000ft in length are not uncommon. All cores should be carefully cataloged and preserved for future inspection. The ratio of core recovery to core length, called the rock quality designation (RQD), is an indicator of rock problems to be encountered.

Groundwater levels should be logged in all borings. Presence of any noxious, explosive, or other gases should be noted.

Where lowering of groundwater may be employed during construction of cut-and-cover or bored tunnels on land, the permeability of the ground should be tested by **pumping tests** in deep wells at selected locations. Rate of pumping and **drawdown** checked in observation wells at various distances should be recorded; aswell as recovery of the water level after stopping the pumps.

Geophysical exploration to determine elevations of distinctive layers of soil or rock surfaces, density, and elastic constants of soil may be used for preliminary investigations. The findings should be verified by a complete boring program before final design and construction.

New Words and Expressions

[1] cut-and-cover construction 盖挖施工；
[2] drill and blast 钻爆；
[3] tunnel boring machine 隧道掘进机；
[4] immersion of prefabricated tunnels 预制隧道浸泡；
[5] sequential excavation methods 连续掘进法；
[6] ground freezing 地面冻结；
[7] tunnel jacking 隧道顶；
[8] critical design loading conditions 关键的设计载荷条件；
[9] appropriate construction methods 适当的施工方法；
[10] clearances for tunnels 隧道间隙；
[11] pedestrian traffic 行人交通；
[12] ventilation [ˌventɪˈleɪʃn] *n.* 通风；
[13] pipeline [ˈpaɪplaɪn] *n.* 管道；
[14] clearances for railroad tunnels 铁路隧道间隙；
[15] clearances for rapid-transit tunnels 快速运输隧道间隙；
[16] rapid-transit tunnels 快速运输通道；
[17] clearances for highway tunnels 公路隧道间隙；
[18] alignment [əˈlaɪnmənt] *n.* 线型；
[19] curvature [ˈkɜːvətʃə(r)] *n.* 曲率；
[20] open lines 明线；
[21] radii of curves 曲线半径；
[22] navigable [ˈnævɪɡəbl] *adj.* & *n.* 通航；可通航的；
[23] cross slope 横坡；
[24] preliminary investigations 初步调查；
[25] topographical features 地形特征；

[26] subsurface structures 地下结构;
[27] seismic ['saɪzmɪk] n. 地震;
[28] reflection [rɪ'flekʃn] n. 反映;
[29] portal ['pɔːtəlz] n. 入口,桥门;
[30] cross section 横截面;
[31] steep dip 陡倾角;
[32] horizontal boring 水平钻孔;
[33] pumping test 扬水试验;
[34] drawdown ['drɔːdaun] n. 水位下降;
[35] geophysical exploration 地球物理勘探。

参考译文:隧道工程

隧道工程使许多重要的水下和地下设施的建设成为可能。为使这些设施的建造者和使用者免受外部环境影响,需要采取独特的设计和施工技术。这些设施必须抵御它们所穿过的物质,包括水在内。它们常常必须承受高压。当用于运输或人类居住时,隧道必须提供足够的照明和安全的空气,并具有去除污染的方法。

隧道的建造有多种方法,取决于隧道经过的土体(岩石)的种类、隧道尺寸与埋深,以及沿路线可能遇到的障碍物。这些方法包括盖挖施工、钻爆施工、隧道掘进机施工、沉管法施工以及顺序开挖法。还有一些特殊施工方法使用频率较低,这些工法常用于非常不利的场地条件,例如冻结法施工和隧道顶管。由于工作时间受限,从高工作压力减压(超过40psi的情况时有发生)需要时间,压缩空气工作方法变得不够经济。劳工组织对压缩空气下工作有相关规定,而且劳工的补偿和健康福利较为昂贵。偶尔仍需在压缩空气条件下工作,例如清理隧道掘进机前面的障碍物,或者对机器的部件进行必要的维护。

地下和水下结构的设计方法与大多数其他结构的设计方法不同。必须首先确定隧道的内部空间、设计寿命和其他要求。其次必须收集地质和环境数据。然后必须确定关键设计荷载条件,包括在极端事件之后隧道的可用条件(例如,隧道要多久才能可重新使用)。然后对适当的施工方法进行评估,以确定最适合的施工方法以满足既定的设备标准、施工条件和成本。永久及临时支护系统与结构本身的施工方法均要考虑。设计标准和操作规范主要适用于地上结构,因此将其应用于地下和水下结构时应注意其适用性。

隧道净空

隧道净空是隧道内表面与隧道中车辆、货物及行人之间的最小距离。将最小净空和隧道中机车车辆限界,以及通风、管线等所需的空间相加可以得到隧道的最小尺寸。

铁路隧道净空。各个铁路有不同的标准,以适应他们的设备。

在铁路隧道中,如果两侧没有壁龛则需留出人员活动距离。在车辆间隙图上,这些间隙至少为每一侧应6英尺8英寸或2米高、30英寸宽,在某些线路上允许最小24英寸。在高速公路隧道中,距离路缘3英尺或0.9米的净距用来作为人行通道。在公路和铁路隧道中,通常在相邻的管道之间沿着共用墙设置一条廊道,以便于两个管道的紧急疏散,避免人员直接进入对面交通流所在的路径。

在弯曲的轨道上,应增加净距以允许长度为85英尺的汽车、60英尺长的卡车的外伸和倾斜，还应保证轨道顶部有 15 英尺 1 英寸的空间。（轨道顶部与系杆顶部之间的距离应为 8 英寸。）

快速运输隧道净空。快速运输隧道的净空设置没有通用标准。要求随系统中使用的车辆大小而异。

公路隧道净空。美国国家公路和运输协会（AASHTO）已为各种类型的高速公路确立了标准的水平和垂直净空。这些标准经过修改已用于联邦公路管理局（FHWA）管辖的州际公路系统。

对于农村和大多数城市的州际公路系统，垂直净空的要求为 16 英尺。

由于隧道的建设成本很高，对净空的要求通常会有所降低。虽然一些较旧的双车道隧道在单向交通路缘之间使用了 21 英尺的道路宽度，对于双向交通使用 23 英尺的道路宽度，并限制车速。然而，这些宽度不再符合 12 英尺或 3.6 米车道的现行标准。由于成本问题路肩很少为全宽，但在每个路缘外至少留 1 英尺。水平曲线周围可能需要更宽的路肩或视距架以符合侧距要求。通常要求墙之间的最小距离为 30 英尺。在没有先移除旧内壁的情况下很少允许在隧道内进行内壁重修，因此对顶部净空不需要预留内壁装修空间。通常在隧道上空中提供车道信号以显示行驶方向上的哪些车道在是开放的，因此需要额外的上部空间，这些空间在适当时也用于布置照明、架立标志、排风扇以及任何其他附在顶棚上的物体。最小竖向交通净空取决于超高车辆使用哪条线路，也取决于高速公路的类别，通常取值介于 14 英尺和 5.1 米之间。线路若有竖向弯曲，还需要增加净空以备长卡车使用。通风、通风设备和通风管道可能需要额外的空间。

隧道的线型及坡度

隧道的水平和垂直的线型通常由曲线连接的直线组成。隧道的最小坡度要保证足够的排水。最大坡度取决于隧道功能。如果可行应尽量在上坡方向修建隧道，因为这样可以保证建设过程中的排水。

铁路隧道的线型和坡度。考虑到列车运行应尽量使用直线和缓坡，但要保证足够的排水，还要考虑建设的总成本。弯曲隧道中的坡度应按曲率进行补偿，就像对于明线所作的一样。一般来说，隧道中的最大坡度不应超过该线路的限制坡度的 75% 左右。该坡度应延伸至隧道下方约 3000 英尺和上方 1000 英尺处。

短而不通风的（低于 2500 英尺）的隧道应在整个长度内坡度不变。长的通风隧道如果从两头开始施工，应在中心设置高点以便在施工期间更好地排水。

曲线半径和路线超高由最大车速控制。

快速交通隧道的线型和坡度。曲率半径和限制坡度由运营条件决定。纽约市 IND 地铁最小半径为 350 英尺，过渡曲线半径小于 2300 英尺。该系统的车站间最大坡度为 3%，而道岔及交叉路口最高坡度为 1.5%。旧金山捷运系统的设计时速为 80 公里。水平曲线半径、轨道超高与速度的关系的关系由下式确定

$$E = \frac{4.65V^2}{R} - U \tag{8.1}$$

式中　E——超高，英寸；

　　　R——半径，英尺；

　　　V——列车速度，英里/小时；

U——不平衡的超高,最好不应超过 $2^{3/4}$ 英寸,或最大不超过 4 英寸。

对于 80 英里/小时的设计速度,具有最优超高的半径为 5000 英尺。对于最大允许超高为 $8^{1/4}$ 英寸的情况,最小半径为 3600 英尺。停车场和道岔的绝对最小半径为 500 英尺。线路最大坡度为 3.0%,车站为 1.0%。为确保良好的排水,坡度应最好不低于 0.50%。

公路线型及坡度。若隧道上方为运输繁忙的通航水域,向上坡度通常限制在 3.5%,向下坡度允许到 4%。若交通量较小,为经济起见也曾使用过 5% 的坡度。在控制通航净空之间,坡度降低到满足排水的最小坡度,纵向不小于 0.25% 并且横向坡度为 1.0%。对于具有双向交通的长岩隧道,最大坡度为 3% 以保证合理的卡车速度。当坡度超过 4% 需增加低速的爬坡车道。

曲率半径应与隧道设计速度相匹配。短半径需要超高和一些拓宽的道路以提供外伸和视距。

初步勘察

应进行勘察以确定所有地形特征,并找到可能受隧道施工影响的所有地表和地下结构。对于水下隧道,应通过测量水深确定河床曲线。

所有隧道施工都必须了解地质条件,对于岩石隧道来说尤为重要。可按需要通过钻孔和地震反射对软土和水下隧道进行勘察。然而对于岩石隧道,特别是长隧道,钻孔的适用性常有限的。应由熟悉该地区的专家进行详细勘察。勘察内容应基于仔细的地表勘察以及对所有可用记录的检查,包括附近其他建设项目记录,例如以前的隧道、矿山、采石场、露天矿、竖井和钻孔。地质学家应该为设计师和承包商出具综合报告。

对于软土地基和水下隧道,钻孔应按一定距离排布。钻孔应该根据当地的条件相隔 500~1000 英尺布置。应在特殊建筑区域使用更紧密的间距,例如通风建筑物、入口和盖挖处。应对土体进行取样和分类,如果可以的话,应对原状样进行实验室检测。实验室不需要的样品,钻孔日志和实验室报告应保留供承包商检查。密度、剪切和压缩强度以及土壤的塑性是最受关注的。

所有钻孔应在隧道仰拱下方进行。对于承压隧道,钻孔应位于隧道横截面之外。

对于岩石隧道,应尽可能多地钻孔。孔可以倾斜以穿过尽可能多的岩层。孔应该在仰拱下方,可能在中心线任意一侧交叉,但应在隧道横截面的外侧,以防止漏水。当隧道通过的地层陡倾,水平钻孔会提供更多信息;钻孔长度为 2000 英尺并不罕见。所有钻芯应仔细分类妥善保存以备将来查验。钻芯收获率与钻芯长度之比称为岩心质量指标(RQD),是常用的岩石质量评价指标。

所有钻孔都应记录地下水位。应注意是否存在任何有毒、易爆或其他气体。

在陆地上采用盖挖施工或钻孔施工时,如果要降低地下水位,应在选定位置的深井中进行抽水试验,以测试场地渗透率。应记录抽水速率和不同距离的观察井的水位下降,还应记录泵停后水位恢复情况。

初步勘察时可以使用地球物理勘探来确定不同土层或岩体分界位置高程、土的密度和弹性模量。在最终设计和施工之前,应通过完整的钻孔程序验证该调查结果。

Text B Tunnel Linings

Unlined Tunnels. Tunnels in very sound rock, not affected by exposure to air, humidity, or freezing, and where appearance is immaterial, are left unlined. This is the case with many railroad tunnels.

Unlined water tunnels in rock are susceptible to leakage either into or out of the tunnel, depending upon the relative pressures. There is therefore a risk that material could be washed out of weak zones and **fissures**, potentially leading to instability unless lined. However, Norwegian hydropower tunnels in good crystalline rock are often unlined for most of their length.

Shotcrete Lining. Where rock is structurally sound but may deteriorate through contact with water or atmospheric conditions, it can be protected by coating with sprayed concrete, reinforced with wire fabric or fibers, or unreinforced.

Such a lining may also be used in water tunnels in good rock to provide a smooth surface, reducing the friction factor and **turbulence.**

Cast-in-Place Concrete. Most tunnels in rock, and all tunnels in softer ground, require a solid lining. Highway tunnels of any importance are always lined for appearance and better lighting conditions. Stone or brick **masonry** has been used to a great extent in the past, but currently concrete is preferred. The thickness of the permanent concrete lining is determined by the size of the tunnel, loading conditions, and the minimum required to embed the steel ribs of any primary lining.

The lining is placed in sections 20 to 30ft long. Segmental steel forms are universally used and must be properly braced to support the weight of the fresh concrete. The walls are usually concreted first, up to the spring line. Next come the arch pours. It is important that the space between the forms and the rock or soil surface be completely filled. Grout pipes should be inserted in the arch concrete to permit filling any voids with sand-and-cement grout.

Concrete is placed through ports in the steel lining or pumped through a pipe introduced in the crown, a so-called slick line. Placement starts at the back of the pour, and the pipe is withdrawn slowly. A combination of both methods may be used. Concrete is either pumped or injected by slugs of compressed air. Admixtures are added to get an easily placed mix with low water content and to reduce concrete **shrinkage.** If there is leakage of water, it usually occurs at shrinkage cracks, which may be sealed with a plastic compound. Or the water may be carried off by copper drainage channels installed in chases cut in the concrete.

Footings for side walls in rock tunnels are cut into the rock below grade. They give adequate stability unless squeezing ground is encountered, in which case a concrete invert lining is placed. In soft ground, a concrete slab is placed, to serve as pavement in highway tunnels. If heavy side pressure exists, this slab may have to be made heavier to prevent **buckling.**

Unreinforced Concrete Lining. A concrete lining is placed to protect the rock and provide a smooth interior surface. Where the concrete lining is exposed to compression stresses only, it may be unreinforced. Most shafts not subject to internal pressure are lined with unreinforced concrete. Shrinkage and temperature cracks are probable and may cause leakage. Where there is a risk of non-uniform loading, unreinforced liners are not used, such as in squeezing ground and through soil overburden.

Reinforced Concrete Lining. In most cases, **reinforcing steel** will be required to withstand tension and bending stresses. Reinforcement is usually required at least on the inside face to resist temperature stresses and shrinkage, although reinforcement elsewhere may be needed to resist moments.

Linings for Shield Tunnels. Linings for shield tunnels may be **one-pass** or two-pass. A one-pass lining system is when the final lining is also the initial lining, usually for tunnels in soil. With a two-pass lining system, an initial lining is installed behind the shield just sufficient to allow the shield to advance while a waterproofing membrane is installed and the final cast-in-place reinforced concrete lining is prepared. The advance rate is thus usually faster and costs fall. The initial lining may be segmental rings with minimal bolting for ease of erection (Fig. 8.1), or steel ribs with lagging. **Precast concrete segments** are now widely used and the use of cast iron and fabricated steel are rare due to their high cost. Although the initial lining may be designed as part of the final lining, any leakage through the seals would result in the full **hydrostatic pressure** acting on the inside final lining for which it should be designed.

Pipe in Tunnel. Water and sewer tunnels up to 14ft diameter are often provided with an internal pipe that forms the inner lining. After the pipe is secured against movement, the space between the initial ground support and the pipe is filled with **cellular** or mass concrete. Sewer pipes may require a further interior lining to protect against corrosive liquids and gases. Water tunnels with a high **internal pressure** exceeding the expected external pressures are usually provided with a steel lining if a reinforced concrete lining is in sufficiently strong. Since the pipe may be **dewatered**, it must also be designed for the **external pressure**, which, if the pipe has leaked, may equal the internal pressure.

In **stiff soils**, steel ribs, usually 4-in H-beams, and wood lagging may be used as primary lining. The ribs are usually spaced 4ft c to c and are erected in the tail of the shield. Precut and dressed wood lagging is placed solidly around the circumference between the flanges of the ribs. This lagging also transfers the jacking forces to the tunnel lining. Precast-concrete lagging has also been used successfully.

Segments are made as long as convenient handling permits, usually 6 to 7ft. The width of the rings depends on the distance the face can be safely excavated ahead of the shield and weight to be handled. The wider the rings, the longer the tail of the shield and hence the more difficult the steering of the shield. Early tunnels had 18-in-wide rings. Recent tunnels have gone to 30 or 32in.

Segments are made to close tolerances on all sides. They are connected by high-strength bolts. Longitudinal joints are offset in successive rings.

The **flanges** have recesses along their matching edges for calking. These **grooves** used to be filled with lead or **impregnated asbestos** calking strips, pounded in manually. Synthetic sealers, such as silicone rubber and polysulfides, can be injected into the grooves by calking guns. These compounds adhere to the metal sufficiently to form an effective seal under pressures usually encountered in underwater tunnels.

Each cast-iron segment is provided with a 2-in grout plug for injection of pea gravel and grout into the space between the lining and the soil. **Bolt holes** are sealed with grommets of impregnated fabric or plastic grommets, the latter being particularly effective. Bolts are tightened with hydraulic or pneumatic wrenches where possible otherwise with hand wrenches.

Welded steel segments, similar in shape to cast-iron segments, have been used for economic reasons in some **subaqueous** tunnels. They were welded in jigs to tolerances as close as practicable, but flanges were not machined and no calking grooves were provided. Difficulties were experienced in making them watertight with **gaskets**. An improved design includes calking grooves and fabrication tolerances similar to those for cast iron.

Precast Concrete. Precast segments are essential to increasing the speed of machine tunneling. A compromise must be reached between the segment size and the number of segments to be installed, directly affecting the weight of the segments, the size of the equipment needed to handle the segments, and the number of operations to be carried out. The width of the segments is governed by the stroke of the jacks pushing the head of the shield, usually in the range of three to five feet. Tapered rings, narrower on one side, are used on bends. At least three segments per ring are required, with five to eight being more common. The closing segment in a ring is usually smaller and wedge shaped to facilitate insertion. Joints in adjacent rings are usually staggered so that all joints are discontinuous, helping to stiffen the rings.

Connection details to adjacent segments vary widely and can be flanged. Straight bolts with nuts, washers and grommets are the most common, but the use of curved recessed bolts result in smaller pockets. Gaining popularity are straight bolts placed at an angle to minimize recesses; the bolts couple into sockets cast into the adjacent section. Dowels may also be used between adjacent rings. The bolts ensure that the rubber or neoprene seals between segments are compressed. The addition of a hydrophilic seal near the outside face may reduce leakage even further. Due to the very close tolerances needed to ensure seals remain watertight and that the diameter remains constant, a high degree of mechanization with steel forms is used. The segments must be installed within the shield tail and the space behind them (the tail void) grouted at a pressure at least equal to the external pressure, making lateral alignment modifications very difficult. It is not uncommon for most bolts to be retrieved once the grout is set. Secondary linings are not essential.

Heavy, **interlocking concrete blocks** have been used successfully in relatively dry or **impervious soil**. They present difficulties when exposed to water pressure due to leakage.

Except where steel rings and lagging or concrete blocks are used as primary lining, no secondary concrete lining is used, unless required for appearance and interior finish of highway tunnels. In this case, a concrete lining of the minimum thickness practicable is placed. When the tunnel is to be faced with tile, provision should be made for attaching it. (To facilitate maintenance and improve lighting, walls and ceilings of highway tunnels are usually finished with ceramic tiles.) To provide good adherence of the scratch coat, scoring wires may be welded longitudinally on the steel forms for the lining to provide a rough concrete surface. Coating of smooth concrete surfaces with epoxy compound may result in satisfactory finishes at less cost.

Design of Tunnel Linings

A liner ring is statically indeterminate. A one-pass lining is designed for transport and erection loads, loads during grouting, and ground loads including seismic. In lieu of computer analyses, which might be as simple as a **two-dimensional** analysis of a grid framework supported on springs, or as complex as finite element or finite difference three-dimensional analyses using **soil-structure interaction** for each step of the construction process, stresses in the liner ring may be computed after the ring is made statically determinate by a cut at the top and one end is fixed (Fig. 8.1).

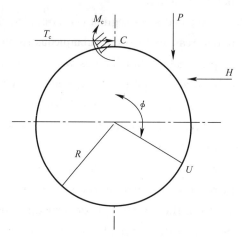

Fig. 8.1 Stresses in liner ring may be computed by assuming it cut at crown C

For a circular ring of constant cross section symmetrically loaded the thrust at the crown C is

$$T_c = \frac{2}{\pi R}\int_0^\pi M\cos\phi\,d\phi \tag{8.2}$$

The vertical shear at the crown is zero, and the moment is

$$M_c = -RT_c - \frac{1}{\pi}\int_0^\pi M\,d\phi \tag{8.3}$$

where R——radius of ring;

   M——bending moment at any point U due to loads on CU;

   ϕ——angle between U and crown C.

With the thrust and moment at the crown known, the stresses at any point on the ring can be computed, as for an arch.

Loads on a lining include its own weight and internal loads, weight of soil above the tunnel (submerged soil for tunnels below water level), reaction due to vertical loads, uniform horizontal pressure due to soil and water above the crown, and triangular horizontal pressure due to soil and water below the crown.

Magnitude of loads on tunnel liners depends on types of soil, depth below surface, loads from adjacent foundations, and surface loads. These will require careful analysis, in which observations made on previous tunnels in similar materials will be most helpful.

In rock, the quality of the rock will affect the loads that are carried by the tunnel, and loads carried by any initial rock support may affect the loads carried by the secondary lining. Compression of **competent rock** due to outward displacement of the tunnel lining in a pressure tunnel may also need to be considered. Often those linings must be designed to take the full internal pressure. Beyond 100 psi internal pressure, reinforced concrete liners may no longer be sufficient and steel liners may be needed. If a tunnel is watertight, the interior lining is usually designed to carry at least the full external water pressure, since leaks in any outer linings will eventually lead to full transfer of the hydraulic head. If the tunnel is drained, at least some of the hydrostatic head should be considered. Blasting may also disturb the rock locally, leading to loads different to those of a bored tunnel.

Following the derivation of moments, **axial thrust and shears**, the concrete cross section can be designed accordingly, and steel or fiber reinforcement placed accordingly as needed. Tension cracks in themselves do not necessarily result in failure, whereas through-cracks (often caused by shrinkage) can cause leakage and corrode exposed steel. It is usually undesirable for cracks to extend more than halfway through the section. Typical steel reinforcement for crack control may reach 0.28% or more of the section area. Restraints at the exterior face due to keying into an irregular rock surface may change the calculated behavior. Linings with irregular width are more likely to crack at the thinnest sections or at initial ground support embedment. Water stops are used at construction joints to reduce leakage.

Because of flexibility, tunnel liner rings can offer only limited resistance to bending produced by unbalanced vertical and horizontal forces. The lining and soil will distort together until a state of equilibrium is obtained. If the deflection, in, exceeds more than $1.5D/10$, where D is the tunnel diameter, ft, the lining may have to be temporarily braced with tie rods when it leaves the shield until the final loading conditions and passive pressures have been developed. In certain soft materials, when shields were **shoved** blind (without material being excavated), initial horizontal pressures exceeded the vertical loads, so that the vertical diameter lengthened temporarily. Ultimately, the section reverted to approximately its initial circular configuration.

When a lining is in rock, determination of the loads imposed on the lining need to be done with care. Stable rock may distribute the stresses around thetunnel, and if impervious, may leave any tunnel lining virtually unloaded. Following excavation, rock that has not yet reached stability can still be moving, extreme examples of which are **squeezing** and **swelling rock**. Further displacements may be little affected by the presence of the lining in such cases, or may depend upon the relative stiffnesses of the two, so that the lining must be designed accordingly. Concrete cast against irregular rock may also be keyed into the rock and result in composite action. If the rock is **anisotropic**, material properties and movements may depend upon direction. Some external loads, such as groundwater pressure and some clays, are independent of displacements. Depending upon the porosity of the surrounding material, water pressure can eventually build up to the full hydrostatic load even in a rock tunnel.

Potable water supply tunnels may need to be made watertight when passing though areas where the inflow of groundwater is not acceptable. Where groundwater contains fine silt or chemicals that could clog drainage facilities on which the tunnel design is based, regular maintenance is required to keep drainage paths clear, or else new drainage paths must be provided or the tunnel designed as watertight. Sewage tunnels frequently generate hydrogen sulfide and so require extra protection against corrosion, such as using an internal PVC or HDPE membrane cast into the internal tunnel lining.

New Words and Expressions
[1] unlined tunnels 无衬砌隧道；
[2] fissure ['fɪʃəz] n. 裂缝；
[3] shotcrete lining 喷射混凝土衬砌；
[4] turbulence ['tɜːbjələns] n. 湍流；
[5] cast-in-place concrete 现浇混凝土；
[6] masonry ['meɪsənri] n. 砌体；
[7] shrinkage ['ʃrɪŋkɪdʒ] v. 收缩，皱缩；
[8] buckling ['bʌklɪŋ] v. 屈曲；
[9] unreinforced concrete lining 无筋混凝土衬砌；
[10] reinforced concrete lining 钢筋混凝土衬砌；
[11] reinforcing steel 钢筋；
[12] linings for shield tunnels 盾构隧道衬砌；
[13] one-pass 一次通过的；
[14] precast concrete segments 预制混凝土砌块；
[15] hydrostatic pressure 静水压力；
[16] pipe in tunnel 管隧道；
[17] water and sewer tunnels 水和污水隧道；
[18] cellular ['seljələ(r)] adj. 多孔的；
[19] internal pressure 内部压力；
[20] dewatered [djuːˈɔːtrd] adj. 脱水的；
[21] external pressure 外部压力；
[22] stiff soils 硬土；
[23] flange [flændʒ] n. (机械等的) 凸缘；
[24] groove [gruːv] n. 沟，槽；
[25] impregnated asbestos 浸渍石棉；
[26] bolt hole 螺栓孔；
[27] subaqueous [sʌbˈeɪkwɪəs] adj. 水下的；

[28] gasket ['gæskɪt] *n.* 垫片；

[29] precast concrete 预制混凝土；

[30] interlocking concrete blocks 嵌锁混凝土砖；

[31] impervious soil 不透水性土壤；

[32] design of tunnel linings 隧道衬砌设计；

[33] two-dimensional 二维，平面；

[34] soil-structureinteraction 土基与建筑物间的相互作用；

[35] competent rock 强岩层；

[36] axial thrust and shears 轴向推力和剪刀；

[37] shove [ʃʌv] *n.* & *v.* 推；

[38] squeezing [sk'wiːzɪŋ] *v.* 挤压；

[39] swelling rock 膨胀岩石；

[40] anisotropic [ænˌaɪsə'trɒpɪk] *adj.* 各向异性的；

[41] potable water 饮用水。

LESSON 9   HIGHWAY ENGINEERING

Text A   Highway Alignments

Geometric design of a highway is concerned with horizontal and vertical alignment. Horizontal alignment of a highway defines its location and orientation in plan view. **Vertical alignment** of a highway deals with its shape in profile. For a roadway with contiguous travel lanes, alignment can be conveniently represented by the centerline of the roadway.

Horizontal Alignment

This comprises one or more of the following geometric elements: **tangents** (straight sections), **circular curves**, and **transition spirals**.

Distance along a horizontal alignment is measured in terms of stations. A full station is defined as 100ft and a half station as 50ft. Station 100+50 is 150ft from the start of the alignment, Station 0+00. A point 1492.27ft from 0+00 is denoted as 14+92.27, indicating a location 14 stations (1400ft) plus 92.27ft from the starting point of the alignment. This distance is measured horizontally along the centerline of the roadway, whether it is a tangent, curve, or a combination of these.

A simple horizontal curve consists of a part of a circle tangent to two straight sections on the horizontal alignment. The radius of a curve preferably should be large enough that drivers do not feel compelled to slow their vehicles. Such a radius, however, is not always feasible, in as much as the alignment should blend harmoniously with the existing topography as much as possible and balance other design considerations, such as overall project economy, sight distance, and side friction. **Superelevation**, usually employed on curves with **sharp curvature**, also should be taken into account.

A curve begins at a point designated **point of curvature** or PC. There, the curve is tangent to the straight section of the alignment, which is called a tangent (Fig. 9.1). The curve ends at **the point of tangency** PT, where the curve is tangent to another straight section of the alignment. The angle Δ formed at PI, **the point of intersection** of the two tangents, is called the **interior angle or intersection angle**.

The curvature of a horizontal alignment can be defined by the radius R of the curve or the degree of curve D. One degree of curve is the central angle that subtends a 100-ft arc (approximately a 100-ft chord). The degree of a curve is given by

$$D = \frac{5729.8}{R} \qquad (9.1)$$

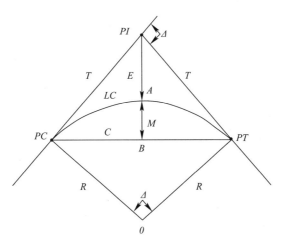

Fig. 9.1 Circular curve starting at point *PC* on one tangent and ending at *PT* on a second tangent that intersects the first one at *PI*. Curve radius is *R* and chord distance between *PC* and *PT* is *C*. Length of arc is *LC*. Tangent distance is *T*

The length of the tangent *T* (distance from *PC* to *PI* or *PI* to *PT*) can be computed from

$$T = R \cdot \tan \frac{\Delta}{2} \tag{9.2}$$

The external distance *E* measured from *PI* to the curve on a radial line is given by

$$E = R \cdot \left(\sec \frac{\Delta}{2} - 1\right) \tag{9.3}$$

The middle ordinate distance *M* extends from the midpoint *B* of the chord to the midpoint *A* of the curve.

$$M = R \cdot \left(1 - \cos \frac{\Delta}{2}\right) \tag{9.4}$$

The length of the chord *C* from *PC* to *PT* is given by

$$C = 2R \cdot \sin \frac{\Delta}{2} = 2T \cdot \cos \frac{\Delta}{2} \tag{9.5}$$

The length *L* of the curve can be computed from

$$L = \frac{\Delta \pi R}{180} = \frac{100\Delta}{D} \tag{9.6}$$

where *D*——intersection angle, degrees.

On starting around a horizontal circular curve, a vehicle and its contents are immediately subjected to centrifugal forces. The faster the vehicle enters the circle and the sharper the curvature, the greater the influence on vehicles and drivers of the change from tangent to curve. For example, depending on the friction between tires and road, vehicles may slide sideways, especially if the road is slick. Furthermore, drivers are uncomfortable because of the difficulty of achieving a position of equilibrium. A similar condition arises when a vehicle leaves a circular curve to enter a straight section of highway. To remedy these conditions, especially where high-speed traffic must round sharp curves, a transition curve with a constantly changing radius should be inserted between the circular curve and the tangent. The radius of the transition curve should vary gradually from infinity at the tangent to that of the circular curve. Along the transition, superelevation should be applied gradually from zero to its full value at the circular curve.

An **Euler spiral** (also known as a clothoid) is typically used as the transition curve. The gradual change in radius results in a corresponding gradual development of centrifugal forces, thereby reducing the aforementioned adverse effects. In general, transition curves are used between tangents and sharp curves and between circular curves of substantially different radii. Transition curves also improve driving safety by making it easier for vehicles to stay in their own lanes on entering or leaving curves. When transition curves are not provided, drivers tend to create their own transition curves by moving laterally within their travel lane and sometimes the adjoining lane, a hazardous maneuver. In addition, transition curves provide a more aesthetically pleasing alignment, giving the highway a smooth appearance without noticeable breaks at the beginning and end of circular curves.

The minimum length L, ft, of a spiral may be computed from

$$L = \frac{3.15V^3}{R \cdot C} \qquad (9.7)$$

where V——vehicle velocity, mi/h;

R——radius, ft, of the circular curve to which the spiral is joined;

C——rate of increase of radial acceleration.

An **empirical value indicative** of the comfort and safety involved, C values often used for highways range from 1 to 3 (For railroads, C is often taken as unity 1).

It is desirable to construct one edge of a roadway higher than the other along curves of highways to counteract centrifugal forces on passengers and vehicles, for the comfort of passengers and to prevent vehicles from overturning or sliding off the road if the centrifugal forces are not counteracted by friction between the roadway and tires. Because of the possibility of vehicle sliding when the curved road is covered with rain, snow, or ice, however, there are limitations on the amount of superelevation that can be used.

The maximum superelevation rate to use depends on local climate and whether the highway is classified as rural or urban. For the safety and comfort of drivers, provision usually is made for gradual change from a tangent to the start of a circular curve. One method for doing this is to insert a spiral curve between those sections of the roadway. A spiral provides a comfortable path for drivers since the radius of curvature of the spiral gradually decreases to that of the circular curve while the superelevation gradually increases from zero to full superelevation of the circular curve. A similar transition is inserted at the end of the circular curve. (An alternative is to utilize compound curves that closely approximate a spiral.) Over the length of the transition, the centerline of each roadway is maintained at profile grade while the outer edge of the roadway is raised and the inner edge is lowered to produce the required superelevation. As indicated in Fig. 9.2, typically the outer edge is raised first until the outer half of the cross section is level with the crown (point B). Then, the outer edge is raised further until the cross section is straight (point C). From there on, the entire cross section is rotated until the full superelevation is attained (point E).

Superelevated roadway cross sections are typically employed on curves of rural highways and urban freeways. Superelevation is rarely used on local streets in residential, commercial, or industrial areas.

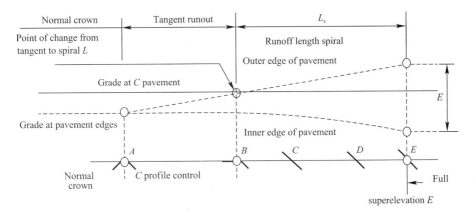

Fig. 9.2 Superelevation variations along a spiral transition curve

A vertical alignment defines the geometry of a highway in elevation, or profile. A vertical alignment can be represented by the highway centerline along a single tangent at a given grade, a vertical curve, or a combination of these.

When a highway is carried on a bridge over an obstruction, a minimum clearance should be maintained between the underside of the bridge superstructure and the feature crossed. AASHTO's (American Association of State Highway and Transportation Officials) Standard Specifications for Highway Bridges specifies an absolute minimum clearance of 14ft and design clearance of 16ft.

These are used as a transition where the vertical alignment changes grade, or slope. Vertical curves are designed to blend as best as possible with the existing topography, consideration being given to the specified design speed, economy, and safety. The tangents to a parabolic curve, known as grades, can affect traffic in many ways; for example, they can influence the speed of large tractor trailers and stopping sight distance.

Although a circular curve can be used for a vertical curve, common practice is to employ a parabolic curve. It is linked to a corresponding horizontal alignment by common stationing. Fig. 9.3 shows a typical vertical curve and its constituent elements.

A curve like the one shown in Fig. 9.3 is known as a crest vertical curve; that is, the curve crests like a hill. If the curve is concave, it is called a sag vertical curve; that is, the curve sags like a valley. As indicated in Fig. 9.3, the transition starts on a tangent at *PVC*, point of vertical curvature, and terminates on a second tangent at *PVT*, point of vertical tangency. The tangents, if extended, would meet at *PVI*. The basic properties of a parabolic vertical curve are derived from an equation of the form $y=ax^2$. The rate of grade change r, percent per station of curve length, is

$$r = \frac{g_2 - g_1}{L} \tag{9.8}$$

where g_1——grade, percent, at *PVC*, shown positive (upward slope) in Fig. 9.3;
g_2——grade, percent, at *PVT*, shown negative (downward slope);
L——length, stations, of vertical curve.

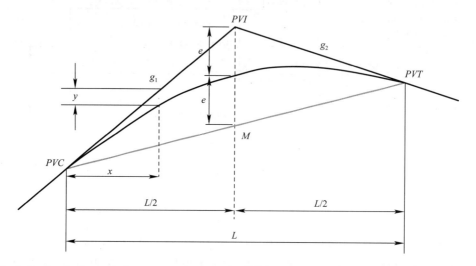

Fig. 9.3 Parabolic vertical curve starting at point PVC on one tangent and terminating at PT on a second tangent that intersects the first one at PVI at a distance e above the curve

If a curve has a length of 700ft, $L=7$. If grade g_1 at *PVC* were 2.25% and grade g_2 at *PVT* were -1.25%, the rate of change would be $r=(-1.25-2.25)/7=-0.50\%$ per station.

A key point on a vertical curve is the turning point, where the minimum or maximum elevation on a vertical curve occurs. The station at this point may be computed from

$$x_{\text{TP}} = \frac{-g_1}{r} \tag{9.9}$$

The middle ordinate distance e, the vertical distance from the *PVI* to the vertical curve, is given by

$$e = \frac{(g_1 - g_2)L}{8} \tag{9.10}$$

For the layout of a vertical curve in the field, it is necessary to know the elevations at points along the curve. From the equation of a parabola, the elevation H_x, ft, of the curve at a distance x, stations, from the *PVC* may be computed from

$$H_x = H_1 + g_1 x + \frac{rx^2}{2} \tag{9.11}$$

where $H_1=$elevation of the *PVC*. The last term of the equation $rx^2/2$ is the vertical offset of the curve from a point on the tangent to the curve at a distance x, stations, from PVC.

New Words and Expressions

［1］ vertical alignment 纵面线形；
［2］ horizontal alignment 平面线形；
［3］ tangent ['tændʒənt] $n.$ 直线区间；
［4］ circular curve 圆曲线；
［5］ transition spiral 螺旋缓和曲线；
［6］ superelevation [sjuːpəelɪ'veɪʃən] $n.$ 超高；
［7］ sharp curvature 曲率；
［8］ point of curvature 曲率点（PC）；
［9］ the point of tangency 切点；
［10］ the point of intersection 交点；
［11］ interior angle or intersection angle 内角和转角；
［12］ euler spiral 欧拉螺线；
［13］ empirical value indicative 经验值表明。

参考译文：公 路 线 形

公路的几何设计包括平面线形和纵面线形。公路平面线形是在平面图上确定它的位置和方向。公路纵面线形指的是从侧面看到的形状。对于一个具有连续行车道的公路，线形可以用路的中心线方便地表示。

平面线形

它包括一项或多项下列几何元素：直线、圆曲线和螺旋曲线。

平面线形上的距离通过桩号来标定。一个整桩 100 英尺，半桩为 50 英尺。100+50 桩号就是从平面线形上的 0+00 桩往前 150 英尺。从 0+00 向前 1492.27 英尺可标注为 14+92.27，表明位置距离平面线形出发点 14 桩加上 92.27 英尺。这个距离是在水平方向沿着中心线测量出来的，不论是直线、曲线还是二者相结合。

平面线形中的一个简单平面曲线可以由连接在一部分圆弧的两条直线段组成。曲线半径需要足够大，以免司机感觉必须降低车速。但是这样的曲线半径并不总能实现，这是由于线型要尽可能与现有地形协调地融合，同时还需平衡其他设计因素，如工程总造价、视距和侧向摩擦力。对于曲率较小的曲线，超高也需被考虑。

一条曲线从指定的曲率点开始（PC 点）。曲线与直线段相切，也称为切线（图 9.1）。曲线在切点 PT 结束，在这曲线与另一条直线相切。PI 处形成的由交点与两条切线构成的角，被称作内角或转角。

平面曲线的曲率可以定义为曲线半径 R 或者曲线角度 D。曲线中的 1 度代表 100-ft 长圆弧所对应的中心角（其弦长约为 100-ft）。曲线角度由下式计算。

$$D = \frac{5729.8}{R} \tag{9.1}$$

直线段长度 T（从 PC 点到 PI 点或 PI 点到 PT 点）可以由下式计算

$$T = R \cdot \tan \frac{\Delta}{2} \tag{9.2}$$

从 PI 点沿半径到曲线的外部距离 E 可以由下式得到

$$E = R \cdot \left(\sec \frac{\Delta}{2} - 1\right) \tag{9.3}$$

从曲线弦中点 B 到曲线中点 A 的矢 M 如下：

$$M = R \cdot \left(1 - \cos \frac{\Delta}{2}\right) \tag{9.4}$$

从 PC 到 PT 点的弦 C 长度如下：

$$C = 2R \cdot \sin \frac{\Delta}{2} = 2T \cdot \cos \frac{\Delta}{2} \tag{9.5}$$

曲线长度 L 由下式计算

$$L = \frac{\Delta \pi R}{180} = \frac{100\Delta}{D} \tag{9.6}$$

式中 D——交角，单位度。

当车辆进入平面圆曲线时，车辆及所载物品会立刻受到离心力作用。进入圆曲线车速越快，曲率越小，直线变到圆对车辆和司机的影响越大。例如，车辆可能产生侧滑，这与轮胎和路面的摩擦有关，如果路面光滑则侧滑更严重。此外，司机很难保持平衡所以感觉难受。当车辆从曲线进入公路直线区域时也会发生类似现象。为减轻这种情况，尤其是高速车流必须通过小曲率曲线时，需要在圆曲线和直线中间插入半径变化的缓和曲线。缓和曲线的半径从直线段的无穷大逐渐变为圆曲线的半径。超高也应沿着缓和曲线从 0 逐渐增加到圆曲线的值。

欧拉螺线（也称为回旋螺线）是典型的缓和曲线。逐渐变化的半径可以产生逐渐增加的离心力，减轻之前提到的有害作用。总体而言，缓和曲线用于直线和小曲率曲线间，以及具有不同半径的圆曲线间。缓和曲线使得车辆在入弯出弯时更容易保持在自身车道中，也改善了行驶安全。当没有缓和曲线时，司机倾向于在车道中自己制造缓和曲线，这是危险行为。此外，缓和曲线提供了更美的线形，使得公路在曲线起始点没有可见的中断，显得更流畅。

螺线的最小长度 L，单位英尺，可以这样计算

$$L = \frac{3.15V^3}{R \cdot C} \tag{9.7}$$

式中 V——车速，单位英里/小时；
R——螺线连接的圆曲线的半径，单位英尺；
C——法向加速度增长率。

由安全性和舒适度的经验表明，公路常用的 C 值为 1~3（铁路常取为 1）。

也需要将道路的一侧比另一侧更高来抵消作用在乘客和车辆上的离心力，这样可以使得乘客更舒服，当轮胎和路面的摩擦力不足以抵消离心力时还可以避免车辆倾覆或侧滑出路面。当道路覆盖有雨水、积雪或冰时车辆可能侧滑，但超高的使用也存在一些限制。

能够使用的最大超高率取决于当地气候以及公路是城市道路还是乡村道路。为了司机的安全舒适，一般在直线段和圆曲线间逐渐施加。一种使用方法是在路的这个位置插入螺

线。螺线的半径逐渐降低到圆曲线，超高逐渐从0增加到圆曲线的超高值，为司机提供了更舒适的路线。在曲线段末端也插入类似的缓和曲线。（另一种方法是用复合曲线来近似螺线）。在螺线的长度上，道路中心线保持竖向高度，道路外侧升高内侧降低以产生所需的超高。如图9.2所示，外侧首先升高到与路顶（B点）平齐。然后，外侧继续升高直到截面变为直线（C点）。然后整个截面旋转直到达到超高值（E点）。

道路超高的典型应用是乡村道路和城市高速。超高很少用于居民区、商业区或工业区的街道。

纵面曲线定义道路在侧向的几何线形。纵面线形可以由公路中心线表示，可以是具有一定斜率的直线，纵向的曲线，或者是二者的结合。

当公路穿越桥梁或其他障碍物时，需要保持距离桥梁结构或其他障碍物下缘的最小净空。AASHTO's（American Association of State Highway and Transportation Officials）标准规定公路桥最小净空为14英尺，设计值为16英尺。

当坡率或倾角发生变化时，这些方法可用于过渡。设计时尽量使得纵面曲线与已有的地形相结合，同时考虑设计车速、经济性和安全性。直线与抛物线相切，又称为坡率，会从多个方面影响交通。例如它们可以影响大型拖拉机的速度，会遮挡视线。

尽管圆曲线可以用于纵面曲线，但实际通常用抛物线。它们与平面线形用公用的桩号联系起来。图9.3显示了一个典型的纵向曲线以及构成的元素。

图9.3所示的曲线称为凸形竖曲线，即曲线如小山一样凸起。如果曲线是凹面的，被称为凹面竖曲线，即曲线如谷地一样下凹。如图9.3所示，曲线从竖向曲率点PVC点开始变化，在第二个竖向切点PVT结束。如果延伸切线会相交于点PVI。竖向抛物线的基本性质可以通过等式$y=ax^2$得到。高度变化率r，即每站间的长度由下式计算

$$r = \frac{g_2 - g_1}{L} \tag{9.8}$$

式中　g_1——PVC处现有坡率，图9.3中以正数表示（上坡）

g_2——PVT处现有坡率，以负数表示（下坡）

L——纵曲线桩间距

如果一个曲线长度为700英尺，$L=7$。如果PVC处的坡率g_1为2.25%，PVT处的坡率g_2为-1.25%，那么桩间的坡率变化率则为$r=(-1.25-2.25)/7=-0.50\%$。

纵曲线的一个关键点是转点，转点处是纵曲线上最大或最小高程处。此点处的桩位可由下述公式计算：

$$x_{TP} = \frac{-g_1}{r} \tag{9.9}$$

纵曲线中点距离e，即PVI点至纵曲线的竖向距离，如下式计算：

$$e = \frac{(g_1 - g_2)L}{8} \tag{9.10}$$

在实际的纵曲线设计中，需要确定曲线上各点的高程。根据抛物线公式，曲线上距离PVC点x处的高程可如下式计算

$$H_x = H_1 + g_1 x + \frac{rx^2}{2} \tag{9.11}$$

式中　H_1——PVC点处的高程。

方程中最后一项 $rx^2/2$ 是距离 PVC 点 x 处切线上一点与曲线在竖直方向的偏差。

Text B Road Surfaces

Roads may be paved with a durable material, such as portland cement concrete or bituminous concrete, or untreated. The economic feasibility of many types of road surfaces depends heavily on the costs and availability locally of suitable materials.

Untreated Road Surfaces

An untreated road surface is one that utilizes untreated soil mixtures composed of gravel, crushed rock, or other locally available material, such as volcanic cinders, blast furnace slag, lime rock, chert, shells, or caliche. Such roads are sometimes used where traffic volume is low, usually no more than about 200 vehicles per day. Should larger traffic volumes develop in the future, the untreated road surface can be used as a subgrade for a higher class of pavement.

To withstand abrasion from superimposed traffic loads, a well-graded coarse aggregate (retained on No. 10 sieve) combined with sand should be used. This mixture provides a tight, water-resistant surface with interlocking aggregate that resists shearing forces. To limit deformation, sufficient binding material, such as clay, may be added to bind the aggregates. Excessive use of clay, however, can lead to surface dislocation brought on by expansion when high moisture is present.

Gravel roads are often used during staged highway construction. Staged construction allows for construction of a project in two or more phases. A dry gravel surface can serve as a temporary road for one phase while construction proceeds on another.

The initial cost of untreated surfaces is often very low compared with that of other types of surfaces. Long-term cost of the roadway may be high, however, because frequent maintenance of the surface may be required. The principal concern in maintenance of untreated road surfaces is providing a smooth surface. Smoothness may be accomplished by blading the surface of the road with a motor grader, drag, or similar device. The roadway cross slopes also need to be maintained; otherwise ponding and other associated drainage problems can occur.

Stabilized Road Surfaces

Controlled mixtures of native soil and an additive, such as asphalt, portland cement, calcium chloride, or sand-clay, can be used to form a stabilized road. Such roads can also serve as a base course for certain types of pavements.

Sand-clay roads are composed of a mixture of clay, silt, fine and coarse sand, and, ideally, some fine gravel. This type of road is frequently used in areas where coarse gravel is not readily available. The thickness of this type of roadway is typically 8in or more. The economic feasibility of sand-clay roads is greatly dependent on the availability of suitable materials.

Calcium chloride ($CaCl_2$) is a white salt with the ability to absorb moisture from the air and then dissolve in the moisture. These properties make it an excellent stabilizing agent and dust palliative. For the latter purpose, calcium chloride is most effective when the surface soil binder is more clayey than sandy.

When calcium chloride is used as a stabilizing agent on an existing surface course, the existing roadway surface should be scarified and mixed with about ½ pound/yd^2 of calcium chloride per inch of depth. For this process to be successful, however, adequate moisture must be present.

The surface of calcium chloride-treated roads is maintained by blading with a motor grader, drag, or similar device. While, under normal conditions, calcium chloride-treated roads generally require less maintenance than untreated surfaces, they require blading immediately after rain. In dry periods, a thin layer of calcium chloride should be applied in order to maintain moisture. During extended dry periods, the road surface may require patching.

Calcium chloride often is used as a deicing agent on pavements and can cause corrosion of the metal bodies of vehicles. Similarly, when used in stabilized roads, calcium chloride can corrode the metal of vehicles, but it also can have adverse environmental effects, such as contamination of groundwater. Accordingly, calcium chloride should be used advisably as a stabilizing and dust control agent.

Untreated road surfaces can be stabilized by mixing the existing road surface with portland cement if the clay content in the soil is favorable for this type of treatment. A general constraint to stabilization with portland cement is that the soils in the road surface contain less than 35% clay. The required rate of application of cement varies with soil classification and generally ranges from 6% to 12% by volume. The roadway surface to be treated should be scarified to accommodate a treated depth of about 6in. The cement should be applied uniformly to the loose material, brought to the optimum moisture content, and then lightly rolled. The quality of soil-cement surfaces can be enhanced by mixing the soils, cement, and water in a central or traveling mixing plant, then rollingthe mixture after ithas been placed on the road.

Various asphalt surface treatments can be utilized to stabilize untreated road surfaces. The process consists of application of asphalt, then aggregate uniformly distributed, and rolling. For double, triple, or other multiple surface treatments, the process is repeated several times. This type of stabilization often is used for roads with low design speeds. Surface treatment with bituminous material should not be expected to accommodate high-speed traffic since vehicles traveling at high speeds tend to dislodge the loose aggregate.

For good results in stabilization with asphalt, at the time of application the temperature should be above 40°F, there should be no rain, and the existing road surface should be dry and well compacted. Also, the quantity and viscosity of the asphalt should be in proper relationship with the temperature, size, and quantity of the aggregate used.

For use as a dust palliative, liquid asphalt may be applied at a rate of 0.1 to 0.5 gal/yd^2. This process is typically referred to as road oiling. This type of dust palliative treatment is often used as a preliminary to progressive improvement of low type roadways.

Flexible Pavements

Bituminous pavements are classified as flexible, whereas portland cement-concrete pavements are considered rigid. Whereas under loads, a rigid pavement acts as a beam that can span across irregularities in an underlying layer, a flexible pavement stays in complete contact with the underlying layer. A rigid pavement is designed so that it can deflect like a beam and then return to the state that existed prior to loading. Flexible pavements, however, may deform and not entirely recover when subjected to repeated loading. The decision as to which type of pavement to use depends on local availability of materials, urban/rural location, traffic volume, costs, and future maintenance considerations.

Fig. 9.4 shows the constituent elements of a typical flexible pavement. The main components, from the bottom up, are the subgrade, subbase, granular base, and asphalt-concrete wearing surface.

Subgrade. This is the underlying soil that serves as the foundation for a flexible pavement. It may be native soil or a layer of selected borrow materials that are compacted to a depth below the surface of the subbase.

Subbase. As shown in Fig. 9.4, the subbase is the course between the subgrade and the base course. The subbase typically consists of a compacted layer of granular material, treated or untreated, or a layer of soil treated with a suitable admixture. It differs from the base course in that it has less stringent specifications for strength, aggregate types, and gradation. If the subgrade meets the requirements of a subbase course, the subbase course may be omitted. In addition to its major structural function as part of the pavement cross section, however, the subbase course can also serve many secondary functions, such as limiting damage due to frost, preventing accumulation of free water within or below the pavement structure, and preventing intrusion of fine-grain subgrade soils into the base courses. In rock cuts, the subbase course can also act as a working platform for construction equipment or for subsequent pavement courses. Performance of these secondary functions depends on the type of material selected for the subbase course.

Base Course. This is the layer of material directly under the surface course. The base course rests on the subbase or, if no subbase is provided, on the subgrade. A structural portion of the pavement, the base course consists of aggregates such as crushed stone, crushed slag, gravel and sand, or a combination of these.

Specifications for base-course materials are much more stringent than those for subbase-course materials. This is especially the case for such properties as strength, stability, hardness, aggregate types, and gradation. Addition of a stabilizing admixture, such as portland cement, asphalt, or lime, can improve the characteristics of a wide variety of materials that, if untreated, would be unsuitable for use as a base course. From an economic standpoint, such treatment is especially beneficial when there is a limited supply of suitable untreated material.

Surface Course. This is the uppermost layer of material in a flexible pavement. It is designed to support anticipated traffic, resist its abrasive forces, limit the amount of surface water that penetrates into the pavement, provide a skid-resistant surface, and offer a smooth riding surface. To serve these purposes, the surface course should be durable, regardless of weather conditions.

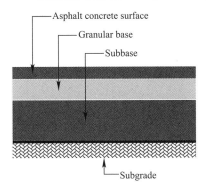

Fig. 9.4 Components of a flexible pavement

Surface courses typically consist of bituminous material and mineral aggregates that are well graded and have a maximum size of about ¾ to 1in. Various other gradations ranging from sand (used in sheet asphalt) to coarse, open-graded mixtures of coarse and fine aggregates have been used with satisfactory results under specific conditions.

Rigid Pavements

A rigid pavement typically consists of a Portland cement-concrete slab resting on a subbase course. (The subbase course may be omitted when the subgrade material is granular.) The slab possesses beamlike characteristics that allow it to span across irregularities in the underlying material. When designed and constructed properly, rigid pavements provide many years of service with relatively low maintenance.

This consists of one or more compacted layers of granular or stabilized material placed between the subgrade and the rigid slab. The subbase provides a uniform, stable, and permanent support for the concrete slab. It also can increase the modulus of subgrade reaction k, reduce or avert the adverse effects of frost, provide a working platform for equipment during construction, and prevent pumping of fine-grain soils at joints, cracks, and edges of the rigid slab.

In design and maintenance of a rigid pavement, a major concern is prevention of accumulation of water on or in the subbase or roadbed soils. Another concern is prevention of erosion, particularly at slab joints and pavement edges. To compensate for this, lean concrete or porous layers are sometimes used as the subbase material. This practice, however, requires close inspection by design and maintenance personnel.

A concrete pavement may be plain concrete, reinforced concrete, or prestressed concrete. Fig. 9.5 shows a cross section of a reinforced concrete pavement. The half cross section in Fig. 9.5 (a) is shown reinforced whereas that in Fig. 9.5 (b) is unreinforced.

Reinforced concrete pavements may be jointed or continuously reinforced. Continuously reinforced pavements eliminate the need for transverse joints but do require construction joints or joints at physical interruptions of the highway, such as bridges. Plain-concrete pavements have no reinforcement except for steel tie bars used to hold longitudinal joints tightly closed.

Jointed Reinforced Concrete Pavement. The main function of reinforcing steel in a jointed concrete pavement is to control cracking caused by thermal expansion and contraction, soil movement, and moisture. The amount and spacing of transverse and longitudinal reinforcing steel required for this purpose depend on slab length, type of steel used, and resistance between the bottom of the slab and the top of the underlying subgrade (or subbase) layer.

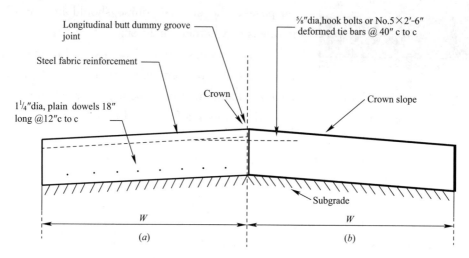

Fig. 9.5　Components of a reinforced concrete pavement

Continuously Reinforced Concrete Pavement. The principal reinforcement in a continuously reinforced pavement is longitudinal steel, which may be reinforcing bars or deformed wire fabric. It is used to control cracking caused by volume changes in the concrete.

In addition to the longitudinal steel, transverse reinforcement may be provided to control the width of longitudinal cracks. When longitudinal cracking is not expected to be troublesome, transverse reinforcement may not be required.

Design of continuously reinforced pavements should take into account the properties of the concrete used. Specifically, the concrete properties that affect the design of continuously reinforced pavements are tensile strength, shrinkage characteristics, and thermal coefficient. Pavement design should also take into account the anticipated drop in temperature, which for design purposes is the difference between the average concrete curing temperature and a design minimum temperature.

New Words and Expressions
[1] untreated road surfaces 未经处理的路面；
[2] stabilized road surfaces 稳定的路面；
[3] flexible pavements 柔性路面；
[4] subgrade ['sʌbgreɪd] n. 路基；
[5] subbase ['sʌbbeɪs] n. 底基层；

［6］base course 基层；

［7］surface course 面层；

［8］rigid pavement 刚性路面；

［9］jointed reinforced concrete pavement 钢筋混凝土路面；

［10］continuously reinforced concrete pavement 连续配筋混凝土路面。

LESSON 10   BRIDGE ENGINEERING

Bridge engineering covers the planning, design, construction, **operation**, and **maintenance of structures** that carry facilities for movement of humans, animals, or materials over natural or created obstacles.

Most of the diagrams used in this section were taken from the "Manual of Bridge Design Practice", State of California Department of Transportation and "Standard Specifications for Highway Bridges", American Association of State Highway and Transportation Officials. The authors express their appreciation for permission to use these illustrations from this comprehensive and authoritative publication.

Bridge Types

Bridges are of two general types: **fixed** and **movable**. They also can be grouped according to the following characteristics:

Supported facilities: Highway or railway bridges and **viaducts, canal bridges** and **aqueducts, pedestrian** or cattle crossings, material-handling bridges, pipeline bridges.

Bridge-over facilities or natural features: Bridges over highways and over railways; river bridges; bay, lake, slough and valley crossings.

Basic geometry: In plan——straight or curved, square or skewed bridges; in elevation——low-level bridges, including causeways and trestles, or highlevel bridges.

Structural systems: Single-span or continuous beam bridges, single or multiple-arch bridges, **suspension bridges**, frame-type bridges.

Construction materials: Timber, **masonry**, concrete, and steel bridges.

Text A (1)   Steel Bridges

Steel is competitive as a construction material for medium and long-span bridges for the following reasons: It has high strength in tension and compression. It behaves as a nearly perfect elastic material within the usual working ranges. It has strength reserves beyond the yield point. The high standards of the fabricating industry guarantee users uniformity of the controlling properties within narrow tolerances. Connection methods are reliable, and workers skilled in their application are available.

The principal disadvantage of steel in bridge construction, its susceptibility to corrosion, is being increasingly overcome by chemical additives or improved protective coatings.

Systems Used for Steel Bridges

The following are typical components of steel bridges. Each may be applied to any of the functional types and structural systems.

Main support: **Rolled beams, plate girders, box girders,** or trusses.

Connections: High-strength bolted, welded, or combinations.

Materials for traffic-carrying deck: Timber stringers and planking, reinforced concrete slab or prestressed concrete slab, stiffened steel plate (orthotropic deck), or steel grid.

Timber decks are restricted to bridges on roads of minor importance. Plates of corrosion resistant steel should be used as ballast supports on through plate-girder bridges for railways. For roadway decks of stiffened steel plates.

Deck framing: Deck resting directly on main members or supported by grids of stringers and floor beams.

Location of deck: On top of main members: deck spans (Fig. 10. 1a); between main members, the underside of the deck framing being flush with that of the main members: through spans (Fig. 10. 1b).

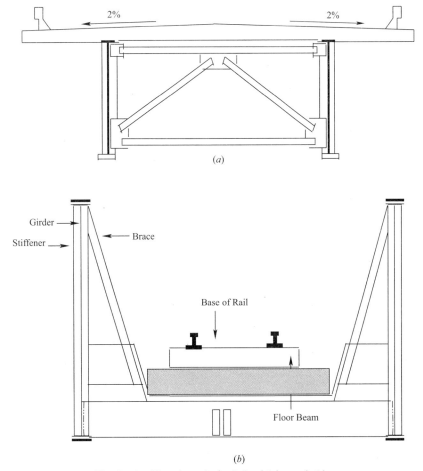

Fig. 10. 1 Two-lane deck-girder highway bridge
(a) Typical section——highway bridge; (b) Typical section——through girder railroad bridge

Rolled-Beam Bridges

The simplest steel bridges consist of rolled **wide-flange** beams and a traffic-carrying deck. Rolled beams serve also as floor beams and stringers for decks of plate-girder and truss bridges.

Reductions in steel weight may be obtained, but with greater labor costs, by adding cover plates in the area of maximum moments, by providing continuity over several spans, by utilizing the deck in composite action, or by a combination of these measures. The principles of design and details are essentially identical with those of plate girders.

Plate-Girder Bridges

The term plate girder applies to structural elements of I-shaped cross section that are welded from plates. Plate girders are used as primary supporting elements in many structural systems: as simple beams on **abutments** or, with overhanging ends, on piers; as continuous or hinged multispan beams; as stiffening girders of arches and suspension bridges, and in frame-type bridges. They also serve as floor beams and stringers on these other bridge systems.

Their prevalent application on highway and railway bridges is in the form of deck-plate girders in combination with concrete decks (Fig. 10.1).

Girders with track ties mounted directly on the **top flanges**, open-deck girders, are used on branch railways and industrial spurs. Through plate girders (Fig. 10.1b) are now practically restricted to railway bridges where allowable structure depth is limited.

The two or more girders supporting each span must be braced against each other to provide stability against overturning and flange buckling, to resist **transverse forces** (wind, earthquake, **centrifugal**), and to distribute concentrated heavy loads. On deck girders, this is done by transverse bracing in vertical planes. Transverse bracing should be installed over each bearing and at intermediate locations not over 25ft apart. This bracing may consist either of full-depth cross frames or of solid diaphragms with depth at least half the web depth for rolled beams and preferably three-quarters the web depth for plate girders. End cross frames or diaphragms should be proportioned to transfer fully all vertical and **lateral loads** to the bearings. On through-girder spans, since top lateral and transverse bracing systems cannot be installed, the top flanges of the girders must be braced against the floor system. For the purpose, heavy **gusset plates** or **knee braces** may be used (Fig. 10.1b).

The most commonly used type of steel bridge girder is the welded plate girder. It is typically laterally braced, non compact, and unsymmetrical, with top and bottom flanges of different sizes.

Variations in moment resistance are obtained by using flange plates of different thicknesses, widths, or steel grades, butt-welded to each other in succession. Web thickness too may be varied. Girder webs should be protected against buckling by transverse and, in the case of deep webs, longitudinal stiffeners. Transverse bearing stiffeners are required to transfer end reactions from the web into the bearings and to introduce concentrated loads into the web. Intermediate and longitudinal stiffeners are required if the girder depth-to-thickness ratios exceed **critical values.**

Composite-Girder Bridges

Installation of appropriately designed **shear connectors** between the top flange of girders or beams and the concrete deck allows use of the deck as part of the top flange (equivalent cover plate). The resulting increase in effective depth of the total section and possible reductions of the top-flange steel usually allow some savings in steel compared with the non composite steel section. The overall economy depends on the cost of the shear connectors and any other additions to the girder or the deck that may be required and on possible limitations in effectiveness of the composite section as such.

In areas of **negative moment**, composite effect may be assumed only if the calculated tensile stresses in the deck are either taken up fully by reinforcing steel or compensated by prestressing. The latter method requires special precautions to assure slipping of the deck on the girder during the prestressing operation but rigidity of connection after completion.

If the steel girder is not shored up while the deck concrete is placed, computation of dead load stresses must be based on the steel section alone.

The effective flange width of the concrete slab that is used as a T-beam flange of a composite girder is the lesser of the following:

1. One-fourth of the span length of the girder;
2. The center to center distance between adjacent girders;
3. Twelve times the least thickness of the slab.

Shear connectors should be capable of resisting all forces tending to separate the abutting concrete and steel surfaces, both horizontally and vertically. Connectors should not obstruct placement and thorough compaction of the concrete. Their installation should not harm the structural steel.

The types of shear connectors presently preferred are channels, or welded studs. Channels should be placed on beam flanges normal to the web and with the channel flanges pointing toward the **girder bearings**.

Orthotropic-Deck Bridges

An **orthotropic deck** is, essentially, a continuous, flat steel plate, with stiffeners (ribs) welded to its underside in a parallel or rectangular pattern. The term orthotropic is shortened from orthogonal **anisotropic**, referring to the mathematical theory used for the flexural analysis of such decks.

When used on steel bridges, orthotropic decks are usually joined quasi-monolithically, by welding or high-strength bolting, to the main girders and floor beams. They then have a dual function as roadway and as structural top flange.

The combination of plate or box girders with orthotropic decks allows the design of bridges of considerable slenderness and of nearly twice the span reached by girders with concrete decks. The most widespread application of orthotropic decks is on continuous, two-to five-span girders on low level river crossings in metropolitan areas, where approaches must be kept short and grades low. This construction has been used for main spans up to 1100ft in **cable-stayed** bridges and up to 856ft without cable stays. There also are some spectacular high-level orthotropic girder bridges and some arch and suspension bridges with orthotropic stiffening girders. On some of the latter, girders and deck have been combined in a single lens-shaped box section that has great stiffness and low **aerodynamic resistance.**

New Words and Expressions

[1] operation [ˌɒpəˈreɪʃn] *n.* 操作，经营；
[2] maintenance of structures 维护结构；
[3] fixed [fɪkst] *adj.* 固定的，不变的；*v.* 固定；
[4] movable [ˈmuːvəbl] *adj.* 活动的，可移动的；*n.* 可移动的东西；
[5] supported facilities 支持设施；
[6] viaduct [ˈvaɪədʌkt] *n.* 高架桥；
[7] canal bridge 管桥；
[8] aqueduct [ˈækwɪdʌkt] *n.* 沟渠；引水渠；导水管；高架渠；
[9] pedestrian [pəˈdestriən] *n.* 步行者；行人；*adj.* 徒步的；平淡的；一般的；
[10] suspension bridge 悬索桥；
[11] masonry [ˈmeɪsənri] *n.* 石工工程，砖瓦工工程；砖石建筑；
[12] rolled beams 轧制梁；
[13] plate girders 板梁；
[14] box girders 箱梁；
[15] wide-flange [ˈwaɪdflˈændʒ] 宽凸缘；
[16] abutment [əˈbʌtmənt] *n.* 桥台；
[17] top flange 上翼缘；
[18] transverse [ˈtrænzvɜːs] *adj.* 横向的；横断的；*n.* 横向物；横轴；横断面；
[19] force [ˈfɔːsɪz] *n.* 力；
[20] centrifugal [ˌsentrɪˈfjuːgl] *adj.* 离心的；
[21] lateral load 横向荷载；
[22] gusset plate 节点板；
[23] knee brace 隅撑；
[24] critical value 临界值；
[25] shear connector 剪力连接件；
[26] negative moment 负弯矩；
[27] girder bearing 梁支座；
[28] orthotropic deck 正交各向异性桥面；
[29] anisotropic [ˌænˌaɪsəˈtrɒpɪk] *adj.* 各向异性的；
[30] cable-stayed 斜拉；
[31] aerodynamic resistance 气动阻力。

参考译文：桥梁工程

桥梁工程包括对运送人、动物或材料跨越自然或人工障碍的结构的规划、设计、建造、运营以及维护。

本节中使用的图表主要摘自加州交通运输部编写的《Manual of Bridge Design Practice（桥梁设计手册）》和美国州立高速公路及运输委员会编写的《Standard Specifications for Highway Bridges（高速公路桥标准化设计）》。能被允许使用这些易理解且权威的著作中的图表，作者深表感谢。

桥的类型

桥梁主要包括两种类型：固定型的和可移动型的。它们也可以根据如下特点进行分类。

按照支撑的设施分：可分为高速公路桥、铁路桥、高架桥、管桥、水渠桥、人行桥、牲畜行桥、材料运送桥、管道桥等。

按桥梁跨越的设施或自然地形分：可分为跨越高速的桥，跨越铁路的桥，跨越河流、海湾、湖泊、沼泽、谷地的桥梁。

按基本几何形状分：从平面上看分为直线桥或曲线桥，直桥或斜桥；从竖向看分为低架桥（包括堤道桥、栈桥）或高架桥。

按结构体系分：可分为单跨桥、连续梁桥、单跨或多跨拱桥、悬索桥、杆桥。

按照建造材料分：可分为木桥、砖石桥、混凝土桥和钢桥。

钢桥

钢材在建设中长跨度的桥梁中比较有竞争力，原因如下：它的拉压强度较高。在通常的工作范围中，它性能接近理想弹性材料。屈服后还保留一定的强度。冶金工业中的高标准确保了使用者可以统一控制材料特性，材料公差小。连接方式可靠，有熟练的技术工人。

钢结构桥梁建造中的主要弊端在于其易于锈蚀，这个弊端正随着化学添加或表面处理而被攻克。

钢结构桥梁体系

如下为典型的钢桥构件。这些构件可能用于任一种功能或结构体系。

主要支撑：轧制梁、板梁、箱梁或桁架。

连接：高强螺栓、焊接或组合连接方式。

桥面支撑材料：木质梁板、混凝土板或预应力混凝土板、加劲性钢板（正交各向异性桥面）或钢格栅。

木桥面仅限于不重要的路桥中使用。耐锈蚀钢板材应用于铁路桥的板梁桥作为道砟支撑。公路桥可用加劲性钢板作为桥面。

桥面系：桥面支撑在主要构件或横纵梁上面。

桥面的位置：在主要结构构件的最上面：桥跨（图10.1a）；在主要构件中间，在桥面系下表面与主要构件齐平：桥跨（图10.1b）。

轧制（钢）梁桥

最简单的钢桥包括宽翼缘钢梁和一个承担交通的桥面。轧制钢梁还可以作为板梁桥和桁架桥中的桥面梁。

在弯矩最大处增加覆板，在几跨间增加连续性，使用桥面的组合作用或者上述方式组合应用可以减小钢的重量，但会增加人工成本。这种设计原理及细节与板梁桥在本质上是一样的。

板梁桥

板梁桥指的是结构构件采用I形截面（译者：焊接工字钢）。焊接工字钢在多种结构体系中：例如作为桥台上的简支梁，或者作为桥墩的悬臂端；或者作为连续的或铰接的多跨桥；或者作为拱桥、悬索桥或杆系类桥的加劲梁。它们也可以作为桥面梁和其他桥型的梁。

它们在公路铁路桥上通常采用梁、桥面加混凝土桥面板的形式（图10.1）。

将轨道直接固定在梁的上翼缘上，采用开放桥面，这用于铁路支线或工业支线。铁路桥的容许的结构宽度有限制，现在下承板梁被限制使用。

支撑每一跨的两个或更多个梁需要支撑在一起以提供抵抗倾覆和翼缘失稳的稳定能力，以及抵抗横向力（风、地震、离心力），并且分配集中在一起的较重的荷载。在桥面梁中，这个是由竖向平面的横撑完成的。横向支撑根据荷载安装，间距不超过25英尺。这种支撑可能包括整个深度的构件，包括深度至少达到腹板深度一半的固体横格（用于轧制钢梁），深度最好达到板梁深度的3/4。端部构件或横格应成比例，将竖向和侧向荷载传递给支撑。在梁跨上，由于侧向支撑无法安装，梁的上翼缘应支撑在桥面上。因此可以采用节点板和隅撑。

应用最多的桥钢梁是焊接钢梁。它典型形式是采用侧向支撑，非紧凑，非对称，上下翼缘尺寸不同。

通过使用对接焊缝将不同厚度、宽度或钢种的翼缘板相连，可以得到不同的抵抗弯矩的能力。腹板厚度也可以不同。梁的腹板需要抵抗失稳，可以通过横向加劲肋或者在深梁中用纵向加劲肋实现。横向支撑加劲肋可以将翼缘上的荷载传导到支座上，可以承受翼缘上的集中荷载。当梁的高厚比超过某一临界值需要加装纵向加劲肋和短加劲肋。

复合梁桥

在梁的上翼缘和混凝土面板中间安装设计好的剪力键，可以使桥面板作为上翼缘（或其他的桥面板）的一部分。这可以增加截面有效高度，与非复合截面的钢桥相比，这可能会减小上翼缘的用钢量。整体的经济水平取决于剪力键和梁或桥面上需要的其他附加装置的成本，还取决于组合截面的有效性的限制条件等。

在负弯矩区域，只有当桥面的拉应力被钢筋或预应力筋抵消的时候才考虑组合效应。后一种方法需要额外注意来保证桥面在施加预应力时可以滑移，但完成后连接应该为刚性。

如果混凝土板放在没有支撑的钢梁上，计算恒荷载应力时只考虑钢的截面。

T梁翼缘有效宽度采用以下较小值：

1. 梁跨长度的1/4；
2. 相邻梁中心距离；
3. 板最小厚度的12倍。

剪力键应该能够抵御导致混凝土和钢表面滑移的所有力，包括水平向和竖向的力。剪力键不应阻碍混凝土浇筑和压实，也不应损坏结构钢。

惯用的剪力键包括槽钢、焊接螺栓。槽钢应该放在梁翼缘上与腹板垂直，槽钢翼缘指向梁支座。

正交异性桥面

正交异性桥面实际上是在钢板下面焊接平行或矩形的加劲肋。正交异性是正交各向异性的简称，指的是用于分析这类桥面弯曲问题的数学理论。

当在钢桥上采用时，正交异性桥面通常由焊缝或高强螺栓将各单片板与主梁或桥面梁连接在一起。它们有两种功能，一个是作为桥面，一个是作为上翼缘。

板梁或箱梁采用正交异性板可以使桥更为纤细，达到混凝土桥面板跨度的两倍。正交异性板桥面最广泛的应用是在城市地区两跨、五跨的连续梁上。这种建造方式已经用于1100英尺的斜拉桥和856英尺的无斜拉桥上。还有一些高等级的梁桥、拱桥和悬索桥也采用正交异性板。在后者中，梁和桥面组合成了透镜形状的箱形截面，这种形式刚度大但气动阻力小。

Text A (2)　Truss Bridges and Suspension Bridges

Truss Bridges

Trusses are lattices formed of straight members in triangular patterns. Although truss-type construction is applicable to practically every static system, the term is restricted here to beam-type structures: simple spans and continuous and **hinged** (cantilever) structures.

Truss bridges require more **field labor** than comparable plate girders. Also, trusses are more costly to maintain because of the more complicated makeup of members and poor accessibility of the exposed steel surfaces. For these reasons, and as a result of changing aesthetic preferences, use of trusses is increasingly restricted to long-span bridges for which the relatively low weight and consequent easier handling of the individual members are decisive advantages.

The superstructure of a typical truss bridge is composed of two main trusses, the **floor system**, a top **lateral system**, a bottom lateral system, cross frames, and **bearing assemblies.**

Decks for highway truss bridges are usually concrete slabs on steel framing. On long-span railway bridges, the tracks are sometimes mounted directly on steel stringers, although continuity of the track ballast across the deck is usually preferred. Orthotropic decks are rarely used on truss bridges.

Most truss bridges have the deck located between the main trusses, with the floor beams framed into the truss posts. As an alternative, the deck framing may be stacked on top of the **top chord**. Deck trusses have the deck at or above top chord level; through trusses, near the **bottom chord.** Through trusses whose depth is insufficient for the installation of a top lateral system are referred to as half through trusses or pony trusses.

Sections of truss members are selected to ensure effective use of material, simple details for connections, and accessibility in fabrication, erection, and maintenance. Preferably, they should be symmetrical.

In bolted design, the members are formed of channels or angles and plates, which are combined into open or half-open sections. Open sides are braced by lacing bars, stay plates, or perforated cover plates. Welded truss members are formed of plates.

The design strength of tensile members is controlled by their net section, that is, by the section area that remains after deduction of rivet or bolt holes. In shop-welded field-bolted construction, it is sometimes economical to build up tensile members by butt-welding three sections of different thickness or steel grades. Thicker plates or higher-strength steel is used for the end sections to compensate for the section loss at the holes.

The **permissible stress** of compression members depends on the **slenderness ratio.** Design specifications also impose restrictions on the width-to-thickness ratios of webs and cover plates to prevent local buckling.

The magnitude of stress variation is restricted for members subject to stress reversal during passage of a moving load.

All built-up members must be stiffened by diaphragms in strategic locations to secure their squareness. Accessibility of all members and connections for fabrication and maintenance should be a primary design consideration.

Whenever possible, each web member should be fabricated in one piece reaching from the top to the bottom chord. The shop length of chord members may extend over several panels. Chord splices should be located near joints and may be incorporated into the gusset plates of a joint.

In most trusses, members are joined by bolting or welding with gusset plates. Pin connections, which were used frequently in earlier truss bridges, are now the exception. As a rule, the centerlines or center-of-gravity lines of all members converging at a joint intersect in a single point.

Stresses in truss members and connections are divided into **primary and secondary stresses.** Primary stresses are the axial stresses in the members of an idealized truss, all of whose joints are made with **frictionless** pins and all of whose loads are applied at pin centers. Secondary stresses are the stresses resulting from the incorrectness of these assumptions. Somewhat higher stresses are allowed when secondary stresses are considered. (Some specifications require computation of the flexural stresses in compression members caused by their own weight as primary stresses.) Under ordinary conditions, secondary stresses must be computed only for members whose depth is more than one-tenth of their length.

Suspension Bridges

These are generally preferred for spans over 1800ft, and they compete with other systems on shorter spans.

The basic structural system consists of flexible main cables and, suspended from them, stiffening girders or trusses (collectively referred to as "stiffening beams"), which carry the deck framing. The vehicular traffic lanes are as a rule accommodated between the main supporting systems. Sidewalks may lie between the main systems or cantilever out on both sides.

Stiffening beams distribute concentrated loads, reduce local deflections, act as chords for the lateral system, and secure the aerodynamic stability of the structure. Spacing of the stiffening beams is controlled by the roadway width but is seldom less than 1/50 the span.

Stiffening beams may be either plate girders, box girders, or trusses. On major bridges, their depth is at least 1/180 of the main span.

The main cables are anchored in massive concrete blocks or, where rock subgrade is capable of resisting **cable tension**, in concrete-filled tunnels. Or the main cables are connected to the ends of the stiffening girders, which then are subjected to longitudinal compression equal to the horizontal component of the cable tension.

Single-span suspension bridges are rare in engineering projects. They may occur in crossings of narrow gorges where the rock on both sides provides a reliable foundation for high-level cable anchorages.

The overwhelming majority of suspension bridges have main cables draped over two towers. Such bridges consist, thus, of a main span and two side spans. Preferred ratios of side span to main span are 1 : 4 to 1 : 2. Ratios of cable sag to main span are preferably in the range of 1 : 9 to 1 : 11, seldom less than 1 : 12.

If the side spans are short enough, the main cables may drop directly from the tower tops to the anchorages, in which case the deck is carried to the abutments on independent, single-span plate girders or trusses. Otherwise, the suspension system is extended over both side spans to the next piers. There, the cables are deflected to the anchorages. The first system allows the designer some latitude in alignment, for example, curved roadways. The second requires straight side spans, in line with the main span. It is the common system for suspension bridges that are links in a chain of multiple-span crossings.

When side spans are not suspended, the stiffening beam is of course restricted to the main span. When side spans are suspended, the stiffening beams of the three spans may be continuous or discontinuous at the towers. The spans are typically restrained to the tower at the ends. Continuity of stiffening beams is required in self-anchored suspension bridges, where the cable ends are anchored to the stiffening beams.

The suspenders between main cables and stiffening beams are usually equally spaced and vertical. Main cables, suspenders, and stiffening beams (girders or trusses) are usually arranged in vertical planes, symmetrical with the longitudinal bridge axis. Bridges with inward-or outward-sloping cables and suspenders and with offset stiffening beams are less common.

Three-dimensional stability is provided by top and bottom lateral systems and transverse frames, similar to those in ordinary girder and truss bridges. Rigid roadway decks may take the place of either or both lateral systems, especially in double decked trusses.

In the United States, the main cables are usually made up of 6-gage **galvanized** bridge wire of 220 to 225ksi ultimate and 82 to maximum 90ksi working stress. The wires are usually placed parallel but sometimes in strands and compacted and wrapped with No. 9 wire. In Europe, strands containing elaborately shaped heat-treated cast-steel wires are sometimes used. Strands must be **prestretched**. They have a lower and less reliable modulus of elasticity than parallel wires. The heaviest cables, those of the Golden Gate Bridge, are about 36in in diameter. Twin cables are used if larger sections are required.

Suspenders may be **eyebars**, **rods**, single steel ropes, or pairs of ropes slung over the main cable. Connections to the main cable are made with cable bands. These are cast steel whose inner faces are molded to fit the main cable. The bands are clamped together with high-strength bolts.

In the design of the floor system, reduction of dead load and resistance to vertical air currents should be the governing considerations. The deck is usually lightweight concrete or steel grating partly filled with concrete with the exception of box sections which usually have a wearing surface. Expansion joints should be provided every 100 to 120ft to prevent **mutual interference** of deck and main structure. **Stringers** should be made composite with the deck for greater strength and stiffness. Floor beams may be plate girders or trusses, depending on available clearance. With trusses, wind resistance is less.

The **towers** may be portal type, multistory, or diagonally braced frames. They may be of **cellular construction**, made of steel plates and shapes, or **steel lattices**, or of reinforced concrete. The substructure below the "spray" line is concrete. The base of steel towers is usually fixed, but it may be hinged. (Hinged towers, however, offer some erection difficulties.) The cable saddles at the top of fixed towers are sometimes placed on rollers to reduce the effect on the towers of unbalanced cable deflections. Cable bents can be considered as short towers, either fixed or hinged, whose axis coincides with the **bisector** of the angle formed by the cable.

New Words and Expressions
[1] hinge [hɪndʒ] *n.* 铰链；
[2] field labor 野外试验室；工地试验室；
[3] floor system 楼面系统；
[4] lateral system 水上航道浮标系统；侧系统；
[5] bearing assembly 轴承组件；
[6] top chord 顶弦，上弦杆；
[7] bottom chord 底弦，下弦；
[8] permissible stress 许用应力；
[9] slenderness ratio 长径比；
[10] primary and secondary stresses 主应力和二次应力；
[11] frictionless [ˈfrɪkʃnles] *adj.* 无摩擦的；
[12] stiffening beam 加劲梁；
[13] cable tension 缆索拉力；
[14] galvanize [ˈgælvənaɪzd] *v.* 用锌镀（铁）；
[15] prestretched 预拉伸；
[16] suspender [səˈspendəz] *n.* 吊杆；吊索；悬挂物；

[17] eyebar [aɪ'bɑː] n. 眼杆，眼铁，带环；
[18] rod ['rɒdz] n. 竿；杆；
[19] mutual interference 相互干扰；
[20] stringer ['strɪŋə(r)] n. 纵梁；
[21] tower ['taʊə(r)] n. 塔；
[22] cellular construction 细胞工程；
[23] steel lattice 钢格；
[24] bisector [baɪ'sektə] n. 平分线。

Text A (3) Cable-Stayed Bridges and Arch Bridges

Cable-Stayed Bridges

The cable-stayed bridge, also called the stayed girder (or truss), has come into wide use since about 1950 for medium-and long-span bridges because of its economy, stiffness, aesthetic qualities, and ease of erection without falsework. Design of cable-stayed bridges utilizes taut cables connecting **pylons** to a span to provide intermediate support for the span.

This principle has been understood by bridge engineers for at least the last two centuries, as indicated by the bridge in Fig. 10.2. The Roeblings used cable stays as supplementary stiffening elements in the famous Brooklyn Bridge (1883). Many recently built and proposed suspension bridges also incorporate taut cable stays when dynamic (railroad) and long-span effects have to be contended with, as in the Salazar Bridge.

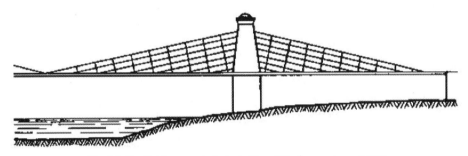

Fig. 10.2 Cable-stayed chain bridge (Hatley system, 1840)

The cable-stayed bridge offers a proper and economical solution for bridge spans intermediate between those suited for deck girders (usually up to 600 to 800 ft but requiring extreme depths, up to 33ft) and the longer-span suspension bridges (over 1000ft). The cable-stayed bridge thus finds application in the general range of 600-to 1600-ft spans but may be competitive in cost for spans as long as 2900ft.

A cable-stayed bridge has the advantage of greater stiffness over a suspension bridge. Use of single or multiple box girders gains large **torsional** and **lateral rigidity**. These factors make the structure stable against wind and aerodynamic effects.

The true action of a cable-stayed bridge (Fig. 10.3) is considerably different from that of a suspension bridge. As contrasted with the relatively flexible cable of the latter, the inclined, taut cables of the cable-stayed structure furnish relatively stable point supports in the main span. Deflections are thus reduced. The structure, in effect, becomes a continuous girder over the piers, with additional intermediate, elastic (yet relatively stiff) supports in the span. As a result, the girder may be shallow. Depths usually range from 1/60 to 1/80 the main span, sometimes even as small as 1/100 the span.

Cable forces are usually balanced between the main and flanking spans, and the structure is internally anchored; that is, it requires no massive masonry anchorages. Analogous to the self anchored suspension bridge, second-order effects of the type requiring analysis by a deflection theory are of relatively minor importance. Thus, static analysis is simpler, and the structural behavior may be more clearly understood.

The above remarks apply to the common, self anchored type of cable-stayed bridges, characterized by compression in the main bridge girders (Fig. 10.3a). It is possible to conceive of the opposite extreme of a fully anchored (earth anchored) cable bridge in which the main girders are in tension. This could be achieved by pinning the girders to the abutments and providing **sliding joints** in the side-span girders adjacent to the pylons (Fig. 10.3b). The fully anchored system is stiffer than the self-anchored system and may be advantageously analyzed by second-order deflection theory because (analogous to suspension bridges) bending moments are reduced by the deformations.

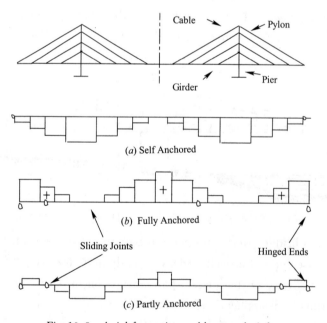

Fig. 10.3 Axial forces in a cable-stayed girder
(a) self-anchored; (b) fully anchored; (c) partly anchored cable-stayed bridges

A further increase in stiffness of the fully anchored system is possible by providing piers in the side spans at the cable attachments (Fig. 10.4). This is advantageous if the side spans are not used for boat traffic below, and if, as is often the case, the side spans cross over low water or land.

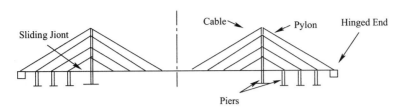

Fig. 10.4 Anchorage of side-span cables at piers and abutments increases stiffness of the center span

A partly anchored cable-stayed system (Fig. 10.3c) has been proposed wherein some of the cables are self-anchored and some fully anchored. The axial forces in the girders are then partly compression and partly tension, but their magnitudes are considerably reduced.

The relatively small diameter of the cables and the absolute minimum amount of overhead structure required are the principal features contributing to the excellent architectural appearance of cable-stayed bridges. The functional character of the structural design produces, as a by-product, a graceful and elegant solution for a bridge crossing. This is encouraged by the wide variety of possible types, using single or multiple cables, including the bundle, harp, fan, and star configurations. These may be symmetrical or asymmetrical.

A wide latitude of choice of cross section of the bridge at the pylons is also possible. The most significant distinction occurs between those with twin pylons (individual, portal, or a frame) and those with single pylons in the center of the roadway. The single pylons usually require a large box girder to resist the torsion of eccentric loadings, and the box is most frequently of steel with an anisotropic steel deck. The single-pylon type is advantageous in allowing a clear unobstructed view from cars passing over the bridge. The pylons may (as with suspension-bridge towers) be either fixed or pinned at their bases. In the case of fixity, this may be either with the girders or directly with the pier.

These structures differ from usual long-span girder bridges in only a few details.

Towers and Floor System The towers are composed basically of two parts: the pier (below the deck) and the pylon (above the deck). The pylons are frequently of steel box cross section, although concrete may also be used.

Bridge Deck Although cable-stiffened bridges usually incorporate an orthotropic steel deck with steel box girders, to reduce the dead load, other types of construction also are in use. For the Lower Yarra River Bridge in Australia, a concrete deck was specified to avoid site welding and to reduce the amount of shop fabrication. The Maracaibo Bridge likewise incorporates a concrete deck, and the Bridge of the Isles (Canada) has a concrete-slab deck supported on longitudinal and transverse steel box girders and steel floor beams. The Buchenauer Bridge also has a concrete deck. Use of a concrete deck in place of orthotropic-plate construction is largely a matter of local economics. The cost of structure to carry the added dead load should be compared with the lower cost per square foot of the concrete deck and other possible advantages, such as better durability and increased stability against wind.

Steel Arch Bridges

A typical arch bridge consists of two or (rarely) more parallel arches or series of arches, plus necessary lateral bracing and end bearings, and columns or hangers for supporting the deck framing. Types of arches correspond roughly to positions of the deck relative to the arch ribs.

Bridges with decks above the arches and clear space underneath (Fig. 10.5a) are designed as open spandrel arches on thrust-resisting abutments. Given enough underclearance and adequate foundations, this type is usually the most economical. Often, it is competitive in cost with other bridge systems.

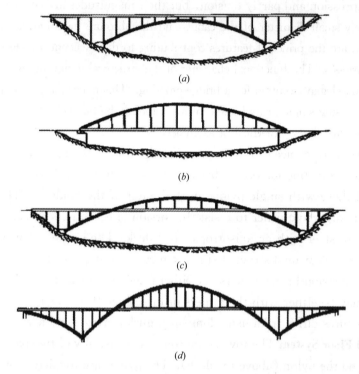

Fig. 10.5 Basic types of steel arch bridges
(a) Open spandrel arch; (b) tied arch; (c) arch with deck at an intermediate level; (d) multiple-arch bridge

Bridges with decks near the level of the arch bearings (Fig. 10.5b) are usually designed as tied arches; that is, tie bars take the arch thrust. End bearings and abutments are similar to those for girder or truss bridges. Tied arches compete in cost with through trusses in locations where underclearances are restricted. Arches sometimes are preferred for aesthetic reasons. Unsightly overhead laterals can be avoided by using arches with sufficiently high moment of **inertia** to resist **buckling**.

Bridges with decks at an intermediate level (Fig. 10.5c) may be tied, may rest on thrustresisting abutments, or may be combined structurally with side spans that **alleviate** the thrust of the main span on the main piers (Fig. 10.5d). Intermediate deck positions are used for long, high-rising spans on low piers.

Spans of multiple-arch bridges are usually structurally separated at the piers. But such bridges may also be designed as continuous structures.

Whether or not **hinges** are required for arch bridges depends on foundation conditions. Abutment movements may sharply increase rib stresses. Fully restrained arches are more sensitive to small abutment movements (and temperature variations) than hinged arches. Flat arches are more sensitive than high arches. If foundations are not fully reliable, hinged bearings should be used.

Complete independence from small abutment movements is achieved by installing a third hinge, usually at the crown. This hinge may be either permanent or temporary during erection, to be locked after all dead-load deformations have been accounted for.

In general, steel arches must be designed for combined stresses due to axial loads and bending.

The **height-to-span ratio** used for steel arches varies within wide limits. Minimum values are around 1 : 10 for tied arches, 1 : 16 for open spandrel arches.

In cross section, steel arches may be **I-shaped, box-shaped**, or tubular. Or they may be designed as space trusses.

The roadway deck of steel arch bridges is usually of reinforced concrete, often of lightweight concrete, on a framing of steel floor beams and stringers. To avoid undesirable co-operation with the primary steel structure, concrete decks either are provided with appropriately spaced expansion joints or prestressed. Orthotropic decks that combine the functions of traffic deck, **tie bar, stiffening girder**, and **lateral diaphragm** have been used on some major arch bridges.

New Words and Expressions

[1] pylon ['paɪlən] *n.* 塔；

[2] torsional ['tɔːʃənəl] *adj.* 扭力的，扭转的；

[3] lateral rigidity 侧向刚度；

[4] sliding joint 滑动接头；

[5] towers and floor system 塔和地板系统；

[6] inertia [ɪ'nɜːʃə] *n.* <物>惯性，惰性；迟钝；不活动；

[7] buckling ['bʌklɪŋ] *n.* 扣住；*v.* 用搭扣扣紧；(使)变形，弯曲；压垮，压弯；

[8] alleviate [ə'liːvieɪt] *vt.* 减轻，缓和；

[9] hinge ['hɪndʒɪz] *n.* 铰链；

[10] height-to-span ratio 高跨比；

[11] I-shaped 工字形；

[12] box-shaped 箱形；

[13] tie bar 拉杆；

[14] stiffening girder 加劲梁；

[15] lateral diaphragm 横隔膜。

Text B (1) Concrete Bridges

Reinforced concrete is used extensively in highway bridges because of its economy in short and medium spans, durability, low maintenance costs, and easy adaptability to horizontal and vertical curvature. The principal types of cast-in-place supporting elements are the longitudinally reinforced slab, T beam or girder, and cellular or box girder. Precast construction, usually prestressed, often employs an I-beam or box-girder cross section. In long-span construction, posttensioned box girders often are used.

Slab Bridges

Concrete slab bridges, longitudinally reinforced, may be simply supported on piers and/or abutments, **monolithic** with wall supports, or continuous over intermediate supports.

The following procedure may be used for design of a typical longitudinally reinforced concrete slab bridge.

Step 1. Determine the live-load distribution (effective width).

Step 2. Assume a slab depth.

Step 3. Determine dead-load moments for the assumed slab depth.

Step 4. Determine live-load moment at point of maximum moment. (This is done at this stage to get a check on the assumed slab depth.)

Step 5. Combine dead-load, live-load, and impact moments at point of maximum moment. Compare the required slab depth with the assumed depth.

Step 6. Adjust the slab depth, if necessary. If the required depth differs from the assumed depth of step 2, the dead-load moments should be revised and step 5 repeated. Usually, the second assumption is sufficient to yield the proper slab depth. Steps 2 through 6 follow conventional structural theory.

Step 7. Place live loads for maximum moments at other points on the structure to obtain intermediate values for drawing envelope curves of maximum moment.

Step 8. Draw the envelope curves. Determine the sizes and points of cutoff for reinforcing bars.

Step 9. Determine distribution steel.

Step 10. Determine the number of piles required at each bent.

Concrete T-Beam Bridges

Widely used in highway construction, this type of bridge consists of a concrete slab supported on, and integral with, girders (Fig. 10.6). It is especially economical in the 50- to 80-ft range. Where falsework is prohibited, because of traffic conditions or clearance limitations, precast construction of reinforced or prestressed concrete may be used.

The structure shown in Fig. 10.6 is a typical fourspan grade-separation structure. The structural frame assumed for analysis is shown in Fig. 10.7. Columns with a pinned base are less stiff than fixed columns which minimizes shrinkage and temperature moments. In addition, foundation pressures in pinned columns are considered fairly uniform, resulting in an economical footing size and design.

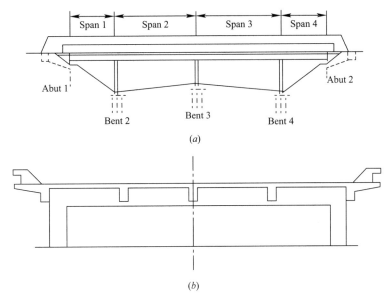

Fig. 10.6 Four-span bridge with concrete T
(a) Elevation; (b) Typical Section

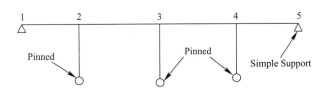

Fig. 10.7 Assumed support conditions for the bridge in Fig. 10.6

For concrete girder design, curves of maximum moments for dead load plus live load plus impact may be developed to determine reinforcement. For live-load moments, truck loadings are moved across the bridge. As they move, they generate changing moments, shears, and reactions. It is necessary to accumulate maximum combinations of moments to provide an adequate design. For heavy moving loads, extensive investigation is necessary to find the maximum stresses in continuous structures.

Concrete Box-Girder Bridges

Box or **hollow** concrete girders are favored by many designers because of the smooth plane of the bottom surface, uncluttered by lines of individual girders. Provision of space in the open cells for utilities is both a structural and an aesthetic advantage. Utilities are supported by the bottom slab, and access can be made available for inspection and repair of utilities.

For sites where structure depth is not severely limited, box girders and T beams have been about equal in price in the 80-ft span range. For shorter spans, T beams usually are cheaper, and for longer spans, box girders. These cost relations hold in general, but box girders have, in some instances, been economical for spans as short as 50ft when structure depth was restricted.

Structural analysis is usually based on two typical segments, interior and exterior girders (Fig. 10.8). An argument could be made for analyzing the entire cross section as a unit because of its inherent transverse stiffness. Requirements in "Standard Specifications for Highway Bridges," American Association of State Highway and Transportation Officials, however, are based on live-load distributions for individual girders, and so design usually is based on the assumption that a box-girder bridge is composed of separate girders.

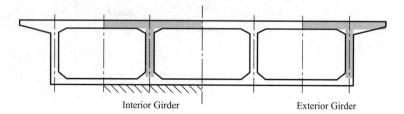

Interior Girder   Exterior Girder

Fig. 10.8 Typical design sections (crosshatched) for a box-girder bridge

Prestressed-Concrete Bridges

In prestressed-concrete construction, concrete is subjected to permanent compressive stresses of such magnitude that little or no tension develops when design loading is applied.

Prestressing allows considerably better utilization of concrete than conventional reinforcement. It results in an overall dead-load reduction, which makes long spans possible with concrete, sometimes competitive in cost with those of steel. Prestressed concrete, however, requires greater sophistication in design, higher quality of materials (both concrete and steel), and greater refinement and controls in fabrication than does reinforced concrete.

Depending on the methods and sequence of fabrication, prestressed concrete may be precast, pretensioned, precast, posttensioned; cast-in-place posttentioned; composite; or partly prestressed.

In precast-beam bridges, the primary structure consists of precast-concrete units, usually I beams, channels, T beams, or box girders. They may be either pretensioned or posttensioned. Precast slabs may be solid or hollow. Precast I beams may be combined with fully or partly cast-in-place decks. This construction has the advantage that the deck can be shaped closely to the desired specifications. Precast slabs, incorporated into the deck, may be used in lieu of removable deck forms where accessibility is poor, for example, on overwater **trestles** or **causeways**. Precast T beams offer no advantage over the easier to fabricate, more compact I sections. Alignment of the flanges of T sections often is difficult. And as with I beams, the flanges must be connected with cast-in-place concrete. Precast box sections may be placed side by side to form a bridge span. If desired, they may be posttensioned transversely.

Precast beams mainly are used for spans up to about 90ft where erection of conventional falsework is not feasible or desirable. Such beams are particularly economical if conditions are favorable for mass fabrication, for example, in multispan **viaducts** or causeways or in the vicinity of centralized fabrication plants. Longer spans are possible but require increasingly heavy handling equipment.

Cast-in-place prestressed concrete often is used for low-level bridges, where ground conditions favor erection of conventional falsework. Typical cross sections are essentially similar to those used for conventionally reinforced sections, except that, in general, prestressing permits structures with thinner depths.

For fully cast-in-place single-span bridges, posttensioning differs only quantitatively from that for precast elements. In design of multispan continuous bridges, the following must be considered: Frictional prestress losses depend on the draping pattern of the ducts. To reduce potential losses and increase the reliability of effective prestress, avoid continuously waving tendon patterns. Instead, use discontinuous simple patterns. Another method is to place tendons, usually bundles of cables, in the hollows of box girders and to bend the tendons at lubricated, accessible bearings.

Prestressed concrete is competitive with other materials for spans of 150 to 250ft or more. Construction techniques and improvements in prestressing hardware, such as smooth, lightweight conduits, which reduce friction losses, have brought prestressed concrete bridges into direct competition with structural steel, once preeminent in medium and long spans.

Segmental construction, both precast and castin-place, has eliminated the need for expensive falsework, which previously made concrete bridges uneconomical in locations requiring long spans over navigation channels or deep canyons. The two types of segmental construction used most in the United States are the cast-in-place and precast balanced **cantilever** types.

For cast-in-place construction, the movable formwork is supported by a structural framework, or traveler, which cantilevers from an adjacent completed section of the superstructure. As each section is cast, cured, and posttensioned, the framework is moved out and the process repeated.

Other methods, such as **full-span** and incremental launching procedures, can be used to fit site conditions. In all segmental construction, special attention should be given in the erection plan to limitation of temporary stresses and to maintenance of balance during erection and prior to span closing. Also important are an accurate prediction of creep and accurate calculation of deflections to ensure attainment of the desired structure profile and deck grades in the completed structure.

New Words and Expressions
[1] monolithic [ˌmɒnə'lɪθɪk] *adj.* 独块巨石的；整体的；庞大的；
[2] hollow ['hɒləʊ] *adj.* 空的；空洞的；空腹的；*n.* 洞，坑；山谷；

[3] trestle ['tresl] *n.* 栈桥；
[4] causeway ['kɔːzweɪ] *n.* 堤道； *v.* 筑堤道于；
[5] viaduct ['vaɪədʌkt] *n.* 高架桥；
[6] segmental construction 节段施工；
[7] cantilever ['kæntɪliːvə(r)] *n.* 悬臂；
[8] full-span 全跨。

Text B (2)   Concrete Bridge Piers and Abutments

Bridge piers are the intermediate supports of the superstructure of bridges with two or more openings. Abutments are the end supports and usually have the additional function of retaining earth fill for the bridge approaches.

The minimum height of piers and abutments is governed by requirements of accessibility for maintenance of the superstructure, including bearings; of protection against spray for bridges over water; and of vertical clearance requirements for bridges over traveled ways. There is no upper limit for pier heights, except that imposed by economic considerations. One of the piers of the Europa Bridge, which carries an international freeway in Austria, for instance, soars to 492ft above the ground surface of the valley.

The top surface of piers must have adequate length and width to accommodate the bridge bearings of the superstructure. On abutments, added width is required for the back wall (curtain wall or **bulkhead**), which retains approach fill and protects the end section of the superstructure. In designing the aboveground sections of piers, restrictions resulting from lateral-clearance requirements of adjacent traveled ways and visibility needs may have to be taken into account. Length and width at the base level are controlled by stability, stress limitations in the pier shaft, and foundation design.

For stress and stability analyses, the reactions from loadings (dead and live, but not **impact**) acting on the superstructure should be combined with those acting directly on the substructure. Longitudinal reactions depend on the type of bearing, whether fixed or expansion.

Piers. A number of basic pier shapes have been developed to meet the widely varying requirements. Enumerated below are some of the more common types and their preferred uses.

Trestle-type piers are preferred on low-level "causeways" carried over shallow waters or seasonally flooded land on concrete slab or beam-and-slab superstructures. Each pier or bent consists of two or more bearing piles, usually all driven in the same plane, and a thick concrete deck or a prismatic cap into which the piles are framed. Both cap and piles may be of timber or, for more permanent construction, of precast conventionally reinforced or prestressed concrete.

Wall-type concrete piers on spread footings are generally used as supports for two-lane overcrossings over divided highways. Given adequate longitudinal support of the superstructure, these piers may be designed as pendulum walls, with joints at top and bottom; otherwise, as cantilever walls.

T-shaped piers on spread footings, with or without bearing piles, may be used to advantage as supports of twin girders. The girders are seated on bearings at both tips of the cross beam atop the pier stem. T-shaped piers have been built either entirely of reinforced concrete or of reinforced concrete in various combinations with structural steel.

Single-column piers of rectangular or circular cross section on spread footings may be used to support box girders, with built-in diaphragms acting as cross beams.

Portal frames may be used as piers under heavy steel girders, with bearings located directly over the portal legs (columns). When more than two girders are to be supported, the designer may choose to strengthen the portal cap beam or to add more columns. Preferably, all legs of each portal frame should rest on a common base plate. If, instead, separate footings are used, as, for instance, on separate pile clusters, adequate tie bars must be used to prevent unintended spreading.

Massive masonry piers have been built since antiquity for multiple-arch river bridges, high-level aqueducts, and more recently, viaducts. In the twentieth century, their place has been taken by massive concrete construction, with or without natural stone facing. Where reduction of dead load is of the essence, hollow piers, often of heavily reinforced concrete, may be used.

Steel towers on concrete pedestals may be used for high bridge piers. They may be designed either as thin-membered, special trellis or as closed box portals, or combinations of these.

Very tall piers, when used, are usually constructed of reinforced or prestressed concrete, either solid or cellular in design.

Bridge abutments basically are piers with flanking (wing) walls. Abutments for short-span concrete bridges, such as T-beam or slab-type highway overcrossings, are frequently simple concrete trestles built monolithically with the superstructure. Abutments for steel bridges and for long-span concrete bridges that are subject to substantial end rotation and longitudinal movements should be designed as separate structures that provide a level area for the bridge bearings (bridge seat) and a back wall (curtain wall or bulkhead). The wall (stem) below the bridge seat of such abutments may be of solid concrete or thin-walled reinforced-concrete construction, with or without counterfort walls; but on rare occasions, masonry is used.

Sidewalls, which retain approach fill, should have adequate length to prevent erosion and undesired spill of the backfill. They may be built either monolithically with the abutment stem and backwall in which case they are designed as cantilevers subject to two-way bending, or as self-supporting retaining walls on independent footings. Sidewalls may be arranged in a straight line with the abutment face, parallel to the bridge axis, or at any intermediate angle to the abutment face that may suit local conditions. Given adequate foundation conditions, the parallel-to-bridge-axis arrangement (U-shaped abutment) is often preferred because of its inherent stability.

Abutments must be safe against overturning about the **toe** of the footing, against sliding on the footing, and against crushing of the underlying soil or overloading of piles. In earth-pressure computations, the vehicular load on highways may be taken into account in the form of an equivalent layer of soil 2ft thick. Live loads from railroads may be assumed to be 0.5kip/ft^2 over a 14-ft-wide strip for each track.

In computations of internal stresses and stability, the weight of the fill material over an inclined or stepped rear face and over reinforced concrete spread footings should be considered as fully effective. No earth pressures however, should be assumed from the earth prism in front of the wall. Buoyancy should be taken into account if it may occur.

New Words and Expressions

[1] bridge pier 桥墩；
[2] bulkhead ['bʌlkhed] *n.* 防水壁；隔风墙；堵墙；
[3] impact ['ɪmpækt] *n.* 撞击；影响；冲击力；*vt.* 挤入，压紧；撞击；对…产生影响；*vi.* 冲撞，冲击；产生影响；
[4] trestle-type pier 栈架式桥墩；
[5] wall-type concrete pier 壁式混凝土桥墩；
[6] T-shaped piers T形墩；
[7] single-column pier 单柱桥墩；
[8] portal frame 门式刚架；
[9] massive masonry pier 大量的圬工桥墩；
[10] steel towers 铁塔；
[11] very tall pier 非常高墩；
[12] toe [təʊ] *n.* 脚趾，脚尖。

LESSON 11 BUILDING ENGINEERING

Text A Structural Systems

Buildings include a wide range of construction intended for human occupancy or for sheltering machines or stored goods. Civil engineers play an important role in the design and construction of such structures. But sometimes, the civil engineer is only one of many design professionals participating in the planning and design of a building. Therefore, it is necessary that the engineer's design decisions take into consideration the objectives and needs of the other professionals. For this purpose, civil engineers must be well-informed on such subjects as architecture, building layout, lighting, electrical systems, elevators, **plumbing**, heating, and air conditioning, as well as structural design. To serve this need, this section summarizes briefly the design principles of those fields and lists references for more detailed study.

Buildings may have load-bearing-wall construction, **skeleton framing**, or a combination of the two. Generally, the engineer's responsibility is to select that type of construction that will serve the owner's total needs most economically. Thus, the most economical construction may not necessarily be the one that requires the least structural materials, or even the one that also has the lowest fabrication and erection costs. Architectural, mechanical, electrical, and other costs that may be affected by the structural system must be considered in any cost comparison.

Because of the large number of variables, which change with time and location, the superiority of one type of construction over the others is difficult to demonstrate, even for a specific building at a given location and time. Availability of materials and familiarity of contractors with required construction methods, or their willingness to take on a job, are important factors that complicate the selection of a structural system still more. Consequently, engineers should consider the specific conditions for each building in selecting the structural system.

Also, deciding on the spans to be used is no simple matter. Foundations, column or wall height, live load, **bracing**, and provisions for ducts and piping vary with each building and must be taken into account, along with the factors previously mentioned. It is possible, however, to standardize designs for simple buildings, such as one-story warehouses or factories, and determine the most economical arrangement and spans of structural components. But such designs should be reviewed and updated periodically because changing conditions, such as the introduction of new materials, new shapes, or new construction methods and price revisions, could change the economic balance.

Engineers also should bear in mind that the relative economics of a structural system can be improved if it can be made to serve more than just structural purposes. Money is saved if a **facade** also carries loads or if a structural slab is both floor and ceiling and also serves as air conditioning ducts.

Load-bearing wood walls frequently are used for one-and two-story houses. They usually consist of 2×4-in studs spaced 24 or 16in c to c and set with wide faces perpendicular to the face of the wall. The walls have top and bottom plates, each consisting of two 2×4's. Unless supported laterally by adequate framing, maximum height of such a wall is 15ft. **Lumber** or **plywood** sheathes the exterior; plaster or wallboard is placed on the interior.

Load-bearing masonry walls have been used for buildings 10 or more stories high. But unless design is based on rational engineering analysis instead of empirical requirements, thickness required at the base is very large. Some building codes require plain masonry bearing walls to have a minimum thickness of 12in for the top 35ft and to increase in thickness 4in for each successive 35ft down. Thus, walls for a 20-story building would have to be about 3ft thick at the bottom.

Since thickness must be increased from the top down, a natural shape in vertical cross section for load-bearing masonry walls is **trapezoidal**. With the widest section at the bottom, such a shape is good for resisting overturning. In practice, however, the **exterior wall face** usually is kept **plumb** and the inside face is stepped where thickness must be increased.

In low buildings, minimum wall thickness may be governed by the ratio of unsupported wall height or length to thickness, whichever ratio is smaller. (For **cavity walls**, thickness is the sum of the **nominal thickness** of inner and outer **wythes**.) Usually, bearing-wall thickness must be at least 6 in; check the local building code.

Much thinner walls can be used with steel-reinforced masonry designed in accordance with Building Code Requirements for "Masonry Structure," Brick Industry Association, Reston, Va.

Load-bearing reinforced concrete walls may be much thinner than masonry for a given height. The American Concrete Institute Building Code Requirements for Reinforced Concrete (ACI 318-86) sets for superstructure walls a minimum thickness of at least 1/25 the unsupported height or length, but not less than 4in. Thickness of exterior basement walls and foundations, however, should be at least $7\frac{1}{2}$ in.

Load-bearing walls may be used for the exterior, partitions, wind bracing, and service-core enclosure. For these purposes, masonry has the disadvantage when used in combinations with skeleton framing of being erected more slowly. Thus, there may be delays in erection of the framing while masonry is being placed to support it.

When load-bearing partitions can be placed at relatively short intervals across the width of a building, curtain walls can be used on the exterior along the length of the building. Such partitions, together with flat-plate reinforced concrete floors (Fig. 11.1), make an efficient structural system for certain types of buildings, such as multistory apartment houses. In such buildings also, concrete walls around closets can double as columns.

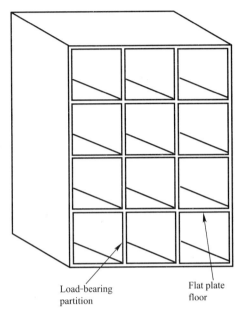

Fig. 11.1 Load-bearing partitions support flat-plate floors in an apartment building

Load-bearing walls may serve as **shear walls**. (But unless they are relatively long, bending stresses due to lateral forces acting parallel to the walls may be large.) Thus, the walls, if properly arranged, will resist wind and earthquake forces in shear and bending.

Load-bearing service-core walls can be designed, however, to carry all the loads in a building. In that case, the roof and floors cantilever from the walls. When spans are large, cantilevers become uneconomical. Instead, columns may assist the service-core walls in carrying the vertical loads. As an alternative, the outer ends of the floors may be suspended from roof trusses, which are supported on but cantilever beyond the core walls. Other possibilities include service cores in pairs with floors supported between them, on girders, trusses, cables, or arches or combinations of these.

Architectural-structural walls represent a type of exterior construction somewhere between load bearing walls and skeleton framing with curtain walls. The load-bearing elements in architectural-structural walls are linear, as in skeleton framing, rather than planar, as in load-bearing walls, and their function is clearly expressed architecturally. Spaces between the structural elements may be screens or curtains, or glass. The structural elements may lie on diagonal lines or verticals; they may be cross-shaped, combining columns and **spandrels**; they may be horizontal or vertical Vierendeel trusses; or they may be any other system that is structurally sound.

In skeleton framing, columns carry building loads to the foundations. Lateral forces are resisted by the columns and diagonal bracing or by rigid-frame action.

Floor and roof construction are much the same for skeleton and load-bearing construction. One principal component is a horizontal structural slab ceiling. The upper surface may serve as or carry a wearing surface for traffic or weather proofing. The deck may be solid, or it may be hollow to reduce weight, permit pipe and wiring to pass through, and serve as air ducts. When the deck does not transmit its loads directly to columns, as it does in flat-slab and flat-plate construction, other major components of floor and roof systems are trusses, beams, and girders (sometimes also called joists, purlins, or rafters, depending on arrangement and location). These support the deck and transmit the load to the columns.

Flat-plate construction employs a deck with constant thickness in each bay and transmits the load directly to columns. It generally is economical forresidential and other lightly loaded structures, where spans are fairly short. It is used for lift-slab construction, in which the **concrete slabs** are cast on the ground, then raised to final position by jacks set on the columns. For longer spans, a waffle or two-way **ribbed plate** may be used.

Flat-slab reinforced concrete construction may be more suitable for heavier loads. Also transmitting loads directly to columns, it differs from flat-plate in that the slab is thickened in the region around the columns (drop panels). Often too, the columns flare at the top (capitals). Waffle construction may be used for longer spans.

Slab-band construction is a variation of flat-plate and flat-slab in which wide, shallow beams are used to support the slab and transmit loads to the columns.

Two-way slabs are another variation; they are supported on girders spanning between columns along the border of each bay. Thus, longer spans and heavier loads can be supported more economically.

Beam-and-girder construction is economical for awide range of conditions. In one-and two-story houses, **wood joists** or **rafters** spaced 16 or 24in c to c generally are used on short spans in conjunction with lumber or plywood decking. For other lightly loaded structures, open-web steel joists, light, rolled-steel beams, or precast-concrete plank may be used, with wood or concrete floors. For heavier loads and longer spans, one-way ribbed-concrete slabs and girders; prestressed-concrete **plank**, tees, double tees, or girders; reinforced concrete beams and girders; laminated-wood girders; or structural-steel beams and girders, including steel-concrete composite construction, may be more suitable. For still longer spans, as usually is the case in industrial buildings, beams and trusses may be most economical.

Arches and catenary construction are appropriate for very long spans. Usually, they are used to support roofs of hangars, stadiums, auditoriums, railroad terminals, and exhibition halls. Their design must provide a means of resisting the horizontal thrust of their reactions.

Thin-shell construction is suitable for uniform loading where curved surfaces are permissible or desirable. It is economical for very long spans. Folded-plate construction often is an economic alternative.

New Words and Expressions

［1］plumbing [ˈplʌmɪŋ] *n.* 管道工程；*v.* 用铅锤测量；探究；
［2］skeleton framing 骨架；
［3］bracing [ˈbreɪsɪŋ] *n.* 紧固（或支撑）装置；*v.* 支住；撑牢；
［4］façade [fəˈsɑːd] *n.* 建筑物的正面，立面；
［5］lumber [ˈlʌmbə(r)] *n.* 木材；
［6］plywood [ˈplaɪwʊd] *n.* 胶合板，合板，夹板；
［7］trapezoidal [træpɪˈzɔɪdəl] *adj.* 梯形的；
［8］exterior wall face 外墙面；
［9］plumb [plʌm] *vt.* 使垂直；用测铅测；探索；*adj.* 垂直的；*n.* 铅锤，测锤；
［10］cavity wall 空心墙；
［11］nominal thickness 额定厚度；
［12］wythes [ˈwɪθs] 烟囱隔板；板层；
［13］shear wall 剪力墙；
［14］spandrel [ˈspændrəl] *n.* 拱肩；
［15］concrete slab 混凝土板；
［16］ribbed plate 肋板；肋形板板；肋形垫板；网纹板；
［17］wood joist 木搁栅；
［18］rafters [ˈræftəz] *n.* 椽；
［19］plank [plæŋk] *n.* （厚）木板；支持物。

参考译文：结 构 系 统

建筑涵盖了人类居住、安置机器及存储物品的多种结构。土木工程师在这种结构的设计和建造中扮演重要角色。但有时土木工程师只是参与建筑规划设计的诸多专业中的一个。因此，工程师在设计时要考虑其他专业的目标和需求。为此，土木工程师除了熟悉结构设计相关知识，还必须熟悉建筑、规划、照明、电气、电梯、给水排水、暖通等专业。为了这项需求，本单元简单总结了这些专业的设计准则，并列举了更细致学习的参考资料。

建筑可能由墙受力、由框架受力或两者结合。通常，工程师的职责就是选择合适建造手段，实现最经济的满足业主需求。因此，最经济的建造方案可能不是最节省建筑材料的，甚至不是施工制造成本最低的。建筑、机械、电气和其他的成本也会受结构体系影响，在比选时要考虑到成本中。

由于建设涉及的参数很多，而且随时间和场地而变化，很难说明一个参数比其他参数更重要，甚至给定建筑的时间和地段也如此。建材是否容易获得，建设单位是否熟悉掌握施工技术，是否有意参与建设，这些重要因素都使得结构体系的选取更为复杂。因此，工程师在选取结构体系时需要具体建筑具体考虑。

对跨度的选择也不是简单的问题。除了之前提到的因素，还要考虑各个建筑不同基础、柱子或墙的高度、活荷载、支撑、管道系统。然而，对于单层仓库、厂房可以确定最经济的样式和结构构件跨度，形成标准化设计。但这种设计必须每隔一段时间就重审升级，因为新材料、新形状、新建造方法的引入以及价格变化等情况都会改变经济性。

工程师脑子里一定要有这样的概念：如果结构可以实现更多目的则经济性更好。如果建筑立面也可以承重，如果结构板既是地板顶棚又是通风管道，则可以节省成本。

用木墙承担荷载常用于一层或两层的房屋中。它们通常包含 2×4 英寸的壁骨，按间距 24 或 16 英寸布置。墙有顶部和底部的板，各包括两块尺寸为 2×4 的板。除非结构有足够的侧向支撑框架，否则墙的最大高度为 15 英尺。可用木板或胶合板做外面装饰，里面是墙板和石膏。

砖墙承重多用于 10 层或更高的建筑。如果没有采用合理的工程结构分析，而采用经验设计，基础需要的厚度会特别大。一些建筑规范规定平面砖墙顶部 35 英尺的厚度为 12 英寸，往下每增加 35 英尺厚度增加 4 英寸。因此，20 层的建筑的底部厚度大约是 3 英尺厚。

因为厚度从上往下逐渐增加，承重砖墙的竖向截面一般是梯形。梯形底部为长边，这有利于抵抗倾覆。然而，实际工程中外墙通常是垂直的，内表面按厚度需要逐渐增加。

在较矮的建筑中，墙厚的最小值通常受无支撑墙高或长厚比的较小值控制（对于空心墙，墙厚是内外板额定厚度的总和）。通常墙厚最小 6 英寸，需要查看当地的建筑规范。

根据 Brick Industry Association，Reston，Va 颁布的砌体建筑规范，用钢加强的砌体的墙的厚度可以选的更薄。

高度相同时，钢混承重墙的厚度要比砖墙薄很多。美国混凝土协会的钢混规范（ACI 318-86）规定上层建筑的最小厚度为无支撑厚度或长度的 25 分之一，但不得低于 4 英寸。地下室外墙及基础厚度最小为 $7\frac{1}{2}$ 英寸。

承重墙也可以用于内墙、隔墙、防风支撑以及设施管道的外封。这时砌体墙与框架合用时存在建造速度慢的缺点。建造支撑框架的砖墙较慢，因此框架的建造也会被拖慢。

当承重的隔墙可以以较小间隔与建筑宽度交叉布置时，建筑长度方向的外墙可以采用屏式管墙。这种隔墙以及钢混的楼板（图 11.1），构成了高效的结构体系，适用于多层公寓结构等建筑。在这种建筑中，房间附近的墙可以加厚作为柱子。

承重墙也可作为剪力墙（但除非墙相对较长，否则平行于墙的侧向力产生的弯曲应力会很大）。因此，这种墙如果合理设计，可以抵御风和地震作用产生的弯矩剪力。

可以设计核心墙承担建筑荷载。在这种情况下，顶棚和楼板从墙中悬伸出来。当跨度较大时，悬伸并不经济。可以用柱子辅助核心墙承担竖向荷载。作为另一种形式，楼板的外端可以悬吊在顶棚桁架上，它受到了核心墙的支撑，但超过核心墙的部分为悬臂构件。其他可能的形式包括将核心墙做成对或组，楼板通过梁、桁架、索、拱或上述构件的组合来支撑在墙中间。

建筑-结构墙是一种介于墙承重和框架承重的建筑形式。建筑-结构墙的承重构件和框架结构是类似线条状的，而不是墙承重中的平面式的，他们的功能在建筑上清晰地表达了。结构构件间的空隙可以用屏风式构件或玻璃填充。结构构件可以呈对角线或竖向布置，它们可能是十字形的，包含柱子和拱券，也可能是采用 Vierendeel 桁架来水平和竖向布置，或采用其他牢固的结构体系。

在框架结构中，柱子将结构荷载传递到基础上。侧向力由柱子和斜向支撑或刚架作用承担。

相对于框架及承重体系建造，楼板和顶棚的建造也基本相同。一个主要的构件是水平向的顶棚。上表面可以作为承担交通或者抵御天气作用的表面。板可以是实心的，也可以是空心的以减轻重量，并让管道、线缆通过。如果板不像在无梁楼板体系中那样直接将荷载传递到柱子上，楼面体系的其他主要构件还包括桁架、梁（有时也称为托梁、檩条、椽等，名称与位置和布置有关）。这些将板的荷载传递到柱子上。

无梁楼盖体系采用各榀厚度相同板，并将板的荷载直接传递到柱子上。它对于居住或其他荷载较小、跨度较小的结构是经济的。它适用于升板施工，这种工法中首先在地面浇筑混凝土板，然后将板用柱上的千斤顶升举到最终位置。对于更长的跨度可采用华夫饼形板或双向肋板。

钢混的无梁楼盖系统更适用于荷载更重的结构。它也将荷载直接传到柱子上。它的不同之处是将柱子附近区域加厚（称为 drop panel）。通常柱子在顶部（柱头）会张开。跨度大时可以使用华夫饼形的板。

Slab-band 工法是无梁楼盖体系的一个变形，它采用宽的镂空的梁来将板的荷载传递到柱子上。

双向板是另一种变形形式。双向板支撑在各榀柱间的梁上。因此，大跨度或荷载较重时这种体系更经济。

梁式体系对于多种情况更经济。在一层或两层的房屋中，中心间隔 16 或 24 英寸的木格栅或椽，多与木板胶合板共同用于小跨度板中。其他荷载较轻的结构，也可采用开口钢檩条、轻轧制钢梁或预制板，并配以木制或混凝土楼板。对于大跨度和重荷载的情况，更合适的结构形式包括单向混凝土肋板和梁、预制板、三通、四通、钢混梁、叠合木板梁、钢组合结构梁（包括钢混凝土组合结构）。对于更大跨度，通常用于工业建筑中的，梁和桁架可能是最经济的。

拱和悬吊体系适用于非常长的跨度。通常它们用于机库、体育馆、礼堂、铁路枢纽和展馆。设计时需要提供抵抗水平向的反力的手段。

薄壁结构适用于荷载均匀、表面可以或需要为曲面的情况。它适用于特别大的跨度。折板式结构是另一种经济的结构形式。

Text B Lateral-Force Bracing

No structural system is complete unless it transmits all forces acting on it into adequate support in the ground. Hence, provision must be made in both low and tall buildings to carry into the foundations not only vertical loads but lateral forces, such as those from wind and earthquake. Also, the possibilities of blast loading and collision with vehicles must be considered. Without adequate provision for resisting lateral forces, buildings may be so unstable that they may collapse during or after construction under loads considerably less than the full design wind or **seismic loads.**

Wood Buildings. In wood-frame houses one and two stories high, plywood or diagonal lumber sheathing may provide adequate resistance to lateral forces if it is properly nailed and glued. With diagonal lumber, each board should be nailed with two nails to every **stud** it crosses. Plywood $\frac{5}{8}$ in thick should be nailed with 8d common nails, 6in c to c; $\frac{1}{4}$ in thick, with 6d nails, 3in c to c. With other types of sheathing, it is advisable to brace the frame with diagonal studs, especially at end corners of the outside walls and important intermediate corners.

Rigid Frames. Buildings of reinforced concrete beam-and-girder construction generally are designed as rigid frames, capable of taking lateral forces. Except possibly for tall structures subjected to severe earthquakes, rarely does additional provision have to be made for bracing against lateral forces. Tall flat-plate buildings also may be designed as rigid frames to resist wind. If the height-width ratio is large, wind resistance can be improved at relatively low cost by placing wings perpendicular or nearly so to the main portion, so that there are rigid frames with several bays parallel to the directions in which wind-force components may be resolved. Thus, the buildings may be made T-shaped, H-shaped, or cross-shaped in plan, or may have V-shaped wings at the ends. Alternatively, buildings may be curved in plan to improve wind resistance.

Shear Walls. When it is impractical to rely on a moment-resisting space frame to take 100% of the lateral forces, shear walls can be used to take all or part of them. Made of structural-steel plates or reinforced brick or concrete, such walls should be long enough parallel to the wind that bending stresses are within the allowable for the concrete and steel. As shown in Fig. 11.2, shear walls may be placed parallel to the narrow width of the building and rigid frames used in the longitudinal direction, or perpendicular shear walls may take lateral forces from any direction, or service-core walls may double as shear walls. The floors should be designed to act as diaphragms or adequate horizontal bracing should be provided to ensure transfer of horizontal forces to the walls. For wind loads, provision must be made to brace exterior walls and transmit the loads from them to the floors. Walls should be adequately anchored to floors and roofs to prevent separation by **wind suction** or seismic forces.

In areas subjected to severe earthquakes, it is advisable that shear walls be supplemented by ductile, moment-resisting space frames, to prevent sudden collapse if the walls should fail.

Braced Frames. Another method of resisting lateral forces is to use diagonal bracing. Frames that are X-braced generally are stiffer than similar frames relying solely on rigid-frame action.

Roof trusses should be braced against horizontal forces since the spans usually are long and **roof decks** are made of light material. Additional horizontal and vertical trusses may be used for the purpose. Also, the framework in the plane of the trusses may be stiffened by inserting knee braces between the columns supporting the trusses and the **bottom chord**. **Purlins** carrying the roof deck should be securely fastened to the top chords, which are in compression, to brace them laterally.

Trussed roof bracing may be placed in the plane of the top or bottom chords. Putting it in the plane of the **top chords** offers the advantages of simpler details, shorter unsupported length of diagonals, and less sagging of bracing because it can be connected to the purlins at all intersections. Bracing both top and bottom chords with separate truss systems seldom is necessary. But the bottom chord should be braced at frequent intervals, even though it is a tension member, to reduce its unsupported length.

Bracing the roof trusses, however, is not enough. The horizontal forces in the roof system must be brought to the ground. The designer must consider the building as a whole.

Tall Buildings. Similarly, when designing bracing for a tall building, the designer should consider the building as a whole. For example, lateral forces may be resisted by all the bents (Fig. 11.2a) or only the outer bents (Fig. 11.2b). In the latter case, the building may be designed as a hollow-tube cantilever for the horizontal forces. The floor and roof systems then must be capable of distributing the loads from the windward wall to the side and **leeward walls.**

For the bents individually, X bracing (Fig. 11.2c) is both efficient and economical. But it usually is impractical because it interferes with doors, windows, and clearance between floors and ceilings above. Generally, the only places X bracing can be installed in tall buildings are in walls without openings, such as elevator-shaft and fire-tower walls. When X bracing cannot be used, bracing that does not interfere with openings should be placed in each bent.

There are many alternatives to X bracing. One is knee bracing between girders and columns (Fig. 11.2d); but the braces may interfere with windows in exterior bents or may be objectionable in interior bents because they are unsightly or reduce floor-to-ceiling clearance. Portal framing of several types, including haunched, solid-web span-drels (Fig. 11.2e) or trusses, are other alternatives. At the columns, these members provide sufficient depth for moment resistance, but at a short distance away from the columns, they become shallow enough to clear windows and doors. In exterior bents, the spandrels can extend from the window head in one story to the window sill in the story above. In interior bents, however, they may have the same disadvantages as knee braces.

Another alternative to diagonal bracing for tall buildings is moment-resistingor wind connections of the bracket type (Fig. 11.2f). Different types may be used, depending on size of members, magnitude of wind moment, and compactness needed to satisfy floor-to-ceiling clearance. In steel framing, the minimum type consists of angles attached to the columns and to top and bottom girder flanges. Plates welded to both girder flanges and butt-welded to the columns are an alternative. When greater moment resistance is needed, the angles may be replaced by tees (made by splitting a wideflange beam at middepth). Also, the bottom flange may be seated on a beam-stub bracket.

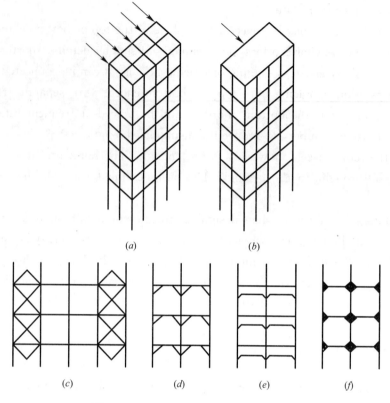

Fig. 11.2 Bracing for high-rise buildings
(a) all transverse bents resist lateral forces; (b) building acts as a vertical tube; (c) bent with X bracing;
(d) knee bracing between columns and girders; (e) haunched spandrels;
(f) moment-resisting connections between columns and girders

New Words and Expressions

[1] seismic load 地震荷载;
[2] wood building 木结构建筑;
[3] stud [stʌd] n. 壁骨; vt. 用壁骨支撑;
[4] rigid frame 钢架;
[5] shear wall 剪力墙;
[6] wind suction 风吸力;
[7] braced frame 支撑框架;
[8] roof truss 屋架;
[9] roof deck 屋顶甲板;
[10] bottom chord 底弦, 下弦;
[11] purlin 檩条;
[12] top chord 上弦;
[13] tall building 高层建筑;
[14] leeward wall 背风墙。

LESSON 12   CONSTRUCTION MANAGEMENT

Text A   Tasks of Construction Management

Construction is the mobilization and utilization of capital and specialized personnel, materials, and equipment to assemble materials and equipment on a specific site in accordance with drawings, specifications, and contract documents prepared to serve the purposes of a client. The organizations that perform construction usually specialize in one of four categories into which construction is usually divided: housing, including single-family homes and apartment buildings; nonresidential building, such as structures erected for institutional, educational, commercial, light-industry, and recreational purposes; engineering construction, which involves works designed by engineers and may be classified as highway construction or heavy construction for bridges, tunnels, railroad, waterways, marine structures, etc.; and industrial construction, such as power plants, steel mills, chemical plants, factories, and other highly technical structures. The reason for such specialization is that construction methods, supervisory skills, labor, and equipment are considerably different for each of the categories.

Construction involves a combination of specialized organizations, engineering science, studied guesses, and calculated risks. It is complex and diversified and the end product typically is nonstandard. Since operations must be performed at the site of the project, often affected by local codes and legal regulations, every project is unique. Furthermore, because of exposure to the outdoors, construction is affected by both daily and seasonal weather variations. It is also often influenced significantly by the availability of local construction financing, labor, materials, and equipment.

Construction Management can be performed by construction contractors, construction consultants also known as construction managers, or design build contractors. All of these individuals or entities have as their goal the most efficient, cost effective completion of a given construction project. Construction contractors typically employ supervisory and **administrative** personnel, labor, materials and equipment to perform construction in accordance with the terms of a contract with a client, or owner. Construction managers may provide guidance to an owner from inception of the project to completion, including oversight of design, approvals, and construction, or just provide construction advisory services to an owner. A construction manager may also act as an agent for the owner, contracting with others for performance of the work and provide administrative and supervisory services during construction. A design build entity can provide all of the above-mentioned activities providing a completed project for the owner with a single contract through one entity.

Construction management can involve the planning, execution, and control of construction operations for any of the aforementioned types of construction.

Planning requires determination of financing methods, estimating of construction costs, scheduling of the work, and selection of construction methods and equipment to be used. Initially, a detailed study of the contract documents is required, leading to compilation of all items of work to be performed and grouping of related items in a master schedule. This is followed by the establishment of a sequence of construction operations. Also, time for execution is allotted for each work item. Subsequent planning steps involve selection of construction methods and equipment to be used for each work item to meet the schedule and minimize construction costs; preparation of a master, or general, construction schedule; development of schedules for **procurement** of labor, materials, and equipment; and forecasts of expenditures and income for the project.

In planning for execution, it is important to recognize that not only construction cost but also the total project cost increases with duration of construction. Hence, fast execution of the work is essential. To achieve this end, construction management must ensure that labor, materials, and equipment are available when needed for the work. Construction management may have the general responsibility for purchasing of materials and equipment and expediting their delivery not only to the job but also to utilization locations. For materials requiring **fabrication** by a supplier, arrangements should be made for preparation and checking of fabrication drawings and inspection of fabrication, if necessary. Also, essential for execution of construction are layout surveys, inspection of construction to check conformance with contract documents, and establishment of measures to ensure job safety and that operations meet Occupational Safety and Health Act (OSHA) regulations and environmental concerns. In addition, successful execution of the work requires provision of temporary construction facilities. These include field offices, access roads, **cofferdams**, drainage, utilities and **sanitation**, and design of formwork for concrete.

Control of construction requires up-to-date information on progress of the work, construction costs, income, and application of measures to correct any of these not meeting forecasts. Progress control typically is based on comparisons of actual performance of construction with forecast performance indicated on master or detailed schedules. Lagging operations generally are speeded by overtime work or addition of more crews and equipment and expedited delivery of materials and equipment to be installed. Cost and income control usually is based on comparisons of actual costs and income with those budgeted at the start of the project. Such comparisons enable discovery of the sources of cost overruns and income shortfalls so that corrective measures can be instituted.

Role of Contractors. The client, or owner, seeking construction of a project, **contracts** with an individual or construction company for performance of all the work and delivery of the finished project within a specific period of time and usually without exceeding estimated cost. This individual or company is referred to as a general contractor.

The general contractor primarily provides construction management for the entire construction process. This contractor may supply forces to perform all of the work, but usually most of the work is subcontracted to others. Nevertheless, the contractor is responsible for all of it. Completely in charge of all field operations, including procurement of construction personnel, materials, and equipment, the contractor **marshals** and allocates these to achieve project completion in the shortest time and at the lowest cost.

The contractor should have two prime objectives: (1) provision to the owner of a service that is satisfactory and on time; (2) making a profit.

Construction Manager. This is a general contractor or construction consultant who performs construction management under a professional service contract with the owner. When engaged at the start of a project, the construction manager will be available to assist the owner and designers by providing information and recommendations on construction technology and economics. The construction manager can also prepare cost estimates during the preliminary design and design development phases, as well as the final cost estimate after completion of the contract documents. Additional tasks include recommending procurement of long-lead-time materials and equipment to ensure **delivery** when needed; review of plans and specifications to avoid conflicts and overlapping in the work of subcontractors; preparing a progress schedule for all project activities of the owner, designers, general contractor, **subcontractors**, and construction manager; and providing all concerned with periodic reports of the status of the job relative to the project schedules. Also, the construction manager, utilizing knowledge of such factors as local labor availability and overlapping trade **jurisdictions**, can offer recommendations concerning the division of work in the specifications that will facilitate bidding and awarding of competitive trade contracts. Furthermore, on behalf of the owner, the manager can take and analyze competitive bids on the work and award or recommend to the owner award of contracts.

During construction, the construction manager may serve as the general contractor or act as anagent of the owner to ensure that the project meets the requirements of the contract documents, legal regulations, and financial obligations. As an agent of the owner, the construction manager assumes the duties of the owner for construction and organizes a staff for the purpose. Other functions of construction management are to provide a resident engineer, or clerk of the works; act as **liaison** with the prime design professional, general contractor, and owner; keep job records; check and report on job progress; direct the general contractor to bring behind-schedule items, if any, up to date; take steps to correct cost overruns, if any; record and authorize with the owner's approval, expenditures and payments; process requests for changes in the work and issue change orders; expedite checking of shop drawings; inspect construction for conformance with contract documents; schedule and conduct job meetings; and perform such other tasks for which an owner would normally be responsible.

New Words and Expressions

[1] administrative [əd'mɪnɪstrətɪv] *adj.* 行政的；
[2] procurement [prə'kjʊəmənt] *n.* 采购；
[3] fabrication [ˌfæbrɪ'keɪʃn] *n.* 制作；
[4] cofferdam ['kɒfədæm] *n.* 围堰；
[5] sanitation [ˌsænɪ'teɪʃn] *n.* 卫生；
[6] role of contractors 承包商的角色；
[7] contract ['kɒntrækt] *n.* 合同；
[8] marshal ['mɑːʃl] *n.* 法警，执法官；
[9] construction manager 施工经理；
[10] delivery [dɪ'lɪvəri] *n.* 交付；
[11] subcontractor [sʌb'kəntræktəz] *n.* 分包商；
[12] jurisdiction [ˌdʒʊərɪs'dɪkʃn] *n.* 司法管辖区；
[13] liaison ['liːeɪzɒn] *n.* 联络。

参考译文：施工管理任务

建设通过对资本、专业技术人员、材料和设备的移动和使用，在特定场地按照预先准备的图纸、要求和合同文件将材料和设备组建成一个结构，满足业主需求。进行施工建设的单位通常专长于如下四类中的一种：（1）房屋，包括独栋别墅、公寓建筑；（2）非住宅结构，包括应用于工业、教育、商业、轻工业和娱乐的建筑；（3）工程结构，包括高速公路、桥梁、隧道、铁路、航道、海事的结构等；（4）工业建筑，包括电厂、钢厂、化工厂、工厂及其他高科技结构。这样划分类别是因为各类别的建设方法、监管技术、劳工、设备均有较大差别。

建设涉及专业的组织、工程科学、研究过的假设和经过计算的风险等的组合。它复杂、多变，最终产品一般不能标准化。由于工程建设必须在工程场地进行，通常还受到当地规范和法律约束，每个工程都是独特的。此外，建设是在室外进行，工程会受到昼夜和季节的天气变化影响。能否得到工程金融、劳动力、材料设备对工程建设有显著影响。

工程管理可以由建造承包方、工程咨询方（也称为工程管理方）或设计承包方执行。这些独立实体都有各自的目标，对于给定工程项目有性价比最高的实现方法。建设承包商通常会雇佣监管和行政人员、劳工、材料、设备等，按照与客户或业主的合同条款进行建造。工程管理方会为业主从工程开始到完工提供咨询，包括设计监督、审查和建设，或者只为业主提供建议服务。工程管理方也可以作为业主的代理，与其他部门签订合同，在建设中提供行政和监理服务。设计或建设实体可以提供上述所有工作，用一个合同为业主提供所有服务。

工程管理也可以包括对上述任何工程建设的规划、执行、运营控制。

规划需要确定采用何种金融方式，评估建设成本，安排工作进度，选择施工方法和使用的设备。开始时需要仔细研究合同文件，汇编工作相关所有事项，并将有关事项在一个总日程表中分组整理。然后建立施工过程的先后顺序。同时给每项工作赋予执行时间。施工顺序规划包括选择施工方法和设备，以满足工程进度并使建设成本最低；准备整个建设进度表；制定人工设备需求计划；以及预测项目收支情况。

在制定执行计划时，需要注意随着工期延长不仅建设成本会增长，项目整体成本也会增长。因此，迅速执行是非常关键的。为此，工程管理需要保证当工作需要时，人工、材料、设备是可用的。施工管理还需要承担如下任务，采购材料、设备并将其运送到工作地点或使用地点。对于需要供货商加工的材料，必要时工程管理需要准备和检查加工图纸，检查加工质量。进行工程建设，本质上要进行测设，检查建造过程确保其满足合同文件，要建立并保证工作安全，符合 Occupational Safety and Health Act（OSHA）规定和满足环保要求。此外，成功的施工工作需要提供临时建造设施。这些包括现场办公室、出入道路、围堰、排水设施和卫生设施，设计混凝土模板。

控制工程建设需要及时更新施工进度、建设成本、收入等信息，如果与预期不符需要采取措施进行纠正。过程控制主要是比较实际工程建设进展与进度总表或详表中预估的建设进展。进度落后需要通过加班或者增加施工人员、设备加快材料和装置的运送安装来加快进度。成本和收入控制通常是比较实际成本和收入，以及工程开始时预算的成本和收入。这种比较可以发现成本超支和收入亏空，以便采取正确措施。

承包商的角色。客户或者业主在寻求项目建设时，要与个人或建设公司订立合同，规定建设质量，规定工程如期完成，不超过预估的成本。这个个人或公司被称为承包商。

总承包商可为整个建设工程提供基本的工程管理。这种承包商可提供工作所需的所有力量，但通常大部分工作会分包给其他人或公司。然而，总承包商负全部责任。承包商管理部门完全负责所有现场运营，包括聘任施工人员、租购材料、设备，分配人员、材料、设备以最短时间最低成本完成项目。

承包商有两个主要目的：（1）按时为业主提供服务，满足业主需求；（2）获取利润。

施工经理。这是一个主要的承包商或建设咨询方，其根据与业主订立的专业服务合同提供施工管理。在工程开始时，施工经理会通过提供信息和提供建造工法和经济性的相关信息来帮助业主和设计方。施工经理还可以在初步设计和详细设计阶段预估成本，在完成合同文件时准备最终成本预估结果。附加任务还包括建议长期采购材料、设备的采购方案，需要时确保货物运输；审查计划、规范以避免分包商的冲突或重叠；准备涉及业主、总承包、分包商、施工经理的所有项目行为的进度计划，按进度计划周期性地为相关方提供工作情况报告。施工经理也可以根据了解的当地劳动力、经销商情况这类信息，提供针对特定工程的建议以使得标书和合同更有竞争力。此外，施工经理可以代表业主收集和分析标书，同时其可进行定标或向业主推荐定标合同。

建造过程中，施工经理可以作为总承包或作为业主代表来确保项目符合合同文件、当地法规和金融要求。作为业主代表，施工经理行使业主责任并组织团队完成这个使命。施工经理的其他作用还包括提供驻场地工程师或项目文书；联络总设计、总承包和业主；做工作记录；检查和回报工作进度情况；如果有进程落后于进度表则指导总承包赶上进度；如果存在成本超支问题则采取措施纠正；记录并批准业主的意见、花销和账单；处理工程中的变更要求，签发变更指令；敦促加工图纸；检查建设情况是否与合同相符；安排并召集工作会议；还包括业主一般需要承担的其他职责。

Text B Organization of Construction Firms

The type of organization employed to carry out construction is influenced by considerations peculiar to that industry, many of which are unlike those affecting manufacturing, **merchandising**, or distribution of goods. This is due largely to the degree of mobility required, type of risk inherent in the particular type of construction, and **geographic** area to be served.

These contracting entities employ the usual business forms. Perhaps the greater number are sole **proprietorships**, where one person owns or controls the enterprise. Many others are partnerships, where two or more individuals form a voluntary association to carry on a business for profit. The **corporate** form has a particular appeal to both large-and small-scale enterprises operating in the construction field. To the large enterprise, corporate structure is an easier way to finance itself by dividing ownership into many small units that can be sold to a wide economic range of purchasers, including those with only small amount of capital to invest. In addition to assisting financing operations, the corporate device brings a limited liability to the persons interested in the enterprise and a **perpetual succession** not affected by the death of any particular owner or by the transfer of any owner's interest. Because of these features, the corporate vehicle is also used by numerous small contractors.

Each facility that a construction team produces, it produces only once; the next time its work will be done at a new location, to a new pattern, and under new, although often similar, specifications. Furthermore, from the very **inception** of each construction project, contractors are wholly devoted to completion of the undertaking as quickly and economically as possible and then moving out.

The problems of construction differ from those of industrial-type businesses. The solutions can best be developed within the construction industry itself, recognizing the unique character of the construction business, which calls for extreme flexibility in its operations. Based on foundations resting within the industry itself, the construction industry has erected organizational structures under which most successful contractors find it necessary to operate. They tend to take executives away from the conference table and put them in close touch with the field. This avoids the type of organizational bureaucracy that hinders rapid communication between office and field and delays vital decisions by management.

Contractor workforces usually are organized by craftsor specialty work classifications. Each unit is directed by a supervisor who reports to a general construction **superintendent** (Fig. 12.1).

The general construction superintendent is in charge of all actual construction, including direction of the production forces, recommendation of construction methods, and selection of personnel, equipment, and materials needed to accomplish the work. This superintendent supervises and coordinates the work of the various craft superintendents and foremen. The general construction superintendent reports to management, or in cases where the magnitude or complexity of the project warrants, to a project manager, who in turn reports to management. To enable the general construction superintendent and project manager to achieve efficient on-the-job production of completed physical facilities, they must be backed up by others not in the direct line of production.

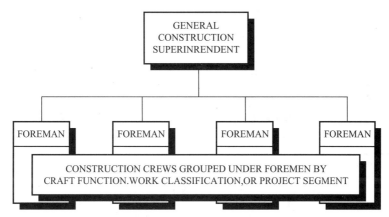

Fig. 12.1 Basic work-performing unit and organization for a small construction company

Fig. 12.1 is representative of the operation of a small contracting business where the sole **proprietor** or owner serves as general construction superintendent. Such owners operate their businesses with limited office help for **payroll** preparation. They may do their own estimating and make commitments for major purchases, but often they use outside accounting and legal services.

As business expands and the owner undertakes larger and more complex jobs, more crafts, functions, or work classifications are involved than can be properly supervised by one person. Accordingly, additional crews with their supervisors may be grouped under as many craft superintendents as required. The latter report to the general construction superintendent, who in turn reports to the project manager, who still may be the owner (Fig. 12.2).

Along with this expansion of field forces, the owner of a one-person business next finds that the volume and complexities of the growing business require specialized support personnel who have to perform such services as

1. Purchasing, receiving, and **warehousing** permanent materials to be incorporated into the completed project, as well as purchasing, receiving, and warehousing goods and supplies consumed or required by the contractor in doing the work.

2. **Timekeeping** and payroll, with all the **ramifications** arising out of federal income tax and Social Security legislation, and detail involved in contracts with organized labor.

3. Accounting and **auditing**, financing, and tax reporting.

4. Engineering estimating, cost control, plant layout, etc.

5. Accident prevention, labor relations, human resources etc.

To coordinate the operation of support staff required for general administration of the business and servicing of its field forces, the head of the organization needs freedom from the direct demands of on-the-job supervision of construction operations. This problem may be solved by employing a general construction superintendent or project manager or by entering into a partnership with an outside person capable of filling that position, with the owner taking the overall management position.

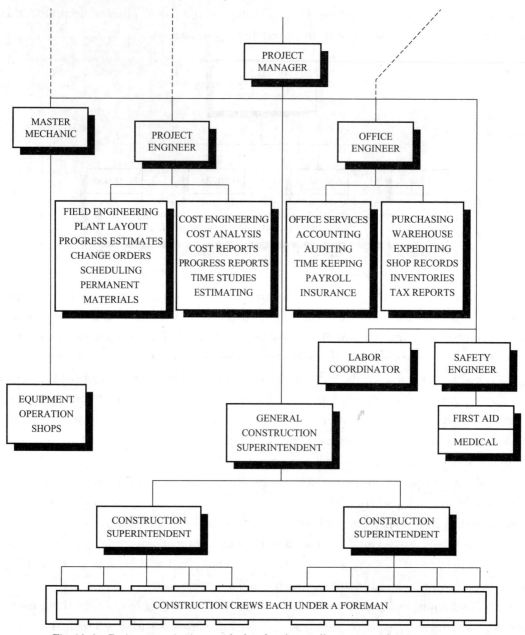

Fig. 12.2 Project organization, with that for the smallest unit as shown in Fig. 12.1

Further growth may find the company operating construction jobs simultaneously at a number of locations. Arrangements for the operation of this type of business take the form of an expanded headquarters organization to administer and control the jobs and service the general construction superintendent or project manager at each location. This concept contemplates, in general, delegation to the field of those duties and responsibilities that cannot best be executed by the headquarters function.

Accordingly, the various jobs usually have a project manager in charge (Fig. 12.2). On small jobs, or in those cases where the general construction superintendent is in direct charge, the project manager is accompanied by service personnel to perform the functions that must be conducted in the field, such as timekeeping, warehousing, and engineering layout.

Some large construction firms, whose operations are regional, nationwide, or worldwide in scope, delegate considerable authority to operate the business to districts or divisions formed on a geographical or functional basis (Fig. 12.3). District managers, themselves frequently corporate officers, are responsible to the general management of the home office for their actions. But they are free to conduct the business within their jurisdiction with less detailed supervision, although within definite confines of well-established company policies. The headquarters office maintains overall administrative control and close communication but constructs projects by and through its district organizations (Fig. 12.4).

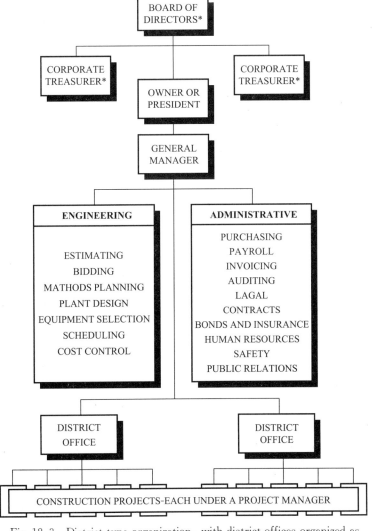

Fig. 12.3 District-type organization, with district offices organized as shown in Fig. 12.4 and projects as indicated in Fig. 12.2

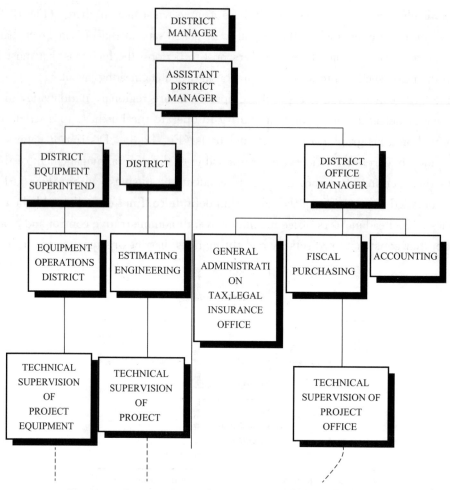

Fig. 12.4 District-type organization for a construction company

Since risk is an important factor in construction, it is only **prudent** to spread it as widely as possible. One safeguard is a joint venture with other contractors whenever the financial hazard of any particular project makes such action **expedient**. In brief, a joint venture is a short-term partnership arrangement wherein each of two or more participating construction companies is committed to a predetermined percentage of a contract and each shares proportionately in the final profit or loss. One of the participating companies acts as the manager or sponsor of the project.

Contractors often employ experts from various disciplines to advise them on conduct of their business. For example, in addition to the usual architectural and engineering consultants, contractors consult the following:

Accountant. Preferably one experienced in construction contracting, the accountant should be familiar with the generally accepted principles of accounting applicable to construction projects, such as costs, actual earnings, and estimated earnings on projects still in progress. Also, the accountant should be able to help formulate the financial status of the contractor, including estimates of the probable earnings from jobs in progress and the amounts of reserves that should be provided for **contingencies** on projects that have been completed but for which final settlements have not been made with all the subcontractors and suppliers.

Attorneys. More than one attorney may be needed to handle a contractor's legal affairs. For example, the contractor may require an attorney for most **routine** matters of corporate business, such as formation of the corporation, registration of the corporation in other states, routine contract advice, and legal aid in general affairs. In addition, the company may need different attorneys to handle claims, personal affairs, estate work, real estate matters, taxes, and dealings with various government bodies.

Insurance and Bonding Brokers. Contractors would be well advised to select an insurance broker who manages a relatively large volume of general insurance. This type of broker can be expected to have large leverage with insurance companies when conditions are encountered involving claims for losses or when influence is needed in establishment of **premiums** at policy renewal time.

For bonding matters, however, contractors will find it advisable to select a broker who specializes in bonding of general contractors and would be helpful in solving their bonding problems. Bonding and general insurance involve entirely different principles. A broker who provides many clients with performance and payment bonds should be able to recommend bonding and insurance companies best suited for the contractor's needs. Also, the broker should be able to assist the contractor and the contractor's accountant in preparation of financial statements with the objective of showing the contractor's position most favorably for bonding purposes.

New Words and Expressions

[1] merchandising ['mɜːtʃəndaɪzɪŋ] *n.* 商品推销；*v.* 推销经营；
[2] geographic [,dʒiːə'ɡræfɪk] *adj.* 地理的；
[3] proprietorship [prə'praɪətəʃɪp] *n.* 独资企业；
[4] corporate ['kɔːpərət] *adj.* 法人的，公司的；
[5] perpetual succession 永久继承；
[6] inception [ɪn'sepʃn] *n.* 成立；开端；
[7] superintendent [,suːpərɪn'tendənt] *n.* 警长；
[8] proprietor [prə'praɪətə(r)] *n.* 业主；
[9] payroll ['peɪrəʊl] *n.* 工资；
[10] warehousing ['weəhaʊzɪŋ] *n.* 仓库费，入仓库；
[11] timekeeping ['taɪmkiːpɪŋ] *n.* 计时；
[12] ramification [,ræmɪfɪ'keɪʃn] *n.* 结果，后果；
[13] audit ['ɔːdɪtɪŋ] *n. & v.* 审计；
[14] prudent ['pruːdnt] *adj.* 谨慎的；
[15] expedient [ɪk'spiːdiənt] *n.* 权宜之计；*v.* 有利的；
[16] accountant [ə'kaʊntənt] *n.* 会计；

[17] contingency [kən'tɪndʒənsɪz] n. 突发事件；
[18] attorney [ə'tɜːni] n. 律师；
[19] insurance and bonding broker 保险和担保的经纪人；
[20] premium ['priːmiəm] n. 保险费；附加费；adj. 高昂的。

Text C Prime Contracts and Subcontracts

A construction contract is an agreement to construct a definite project in accordance with plans and specifications for an agreed sum and to complete it, ready for use and occupancy, within a certain time. Although contractsmay be expressed or implied, oral or written, agreements between owners and contractors are almost universally reduced to writing. Their forms may vary from the simple acceptance of an offer to the usual fully documented contracts in which the complete plans, specifications, and other instruments used in bidding, including the contractor's proposal, are made a part of the contract by reference.

Recognizing that there are advantages to standardization and simplification of construction contracts, the Joint Conference on Standard Construction Contracts prepared standard documents for construction contracts intended to be fair to both parties. The American Institute of Architects also has developed standard contract documents. And the Contract Committees of the American Society of Municipal Engineers and the Associated General Contractors of America have proposed and approved a Standard Code for Municipal Construction.

Contractors generally secure business by submitting proposals in response to invitations to bid or by negotiations initiated by either party without formal invitation or competitive bidding. Agencies and **instrumentalities** of the federal government and most state and municipal governments, however, are generally required by law to let construction contracts only on the basis of competitive bidding. However, certain federal agencies, for security reasons or in an emergency, may restrict bidders to a selected list, and, in these cases, may not open bids in public.

Normally, competitive bidding leads to fixed-price contracts. These may set either a lump-sum price for the job as a whole or unit prices to be paid for the number of prescribed units of work actually performed. Although negotiated contracts may be on a lump-sum or unit-price basis, they often take other forms embodying devices for making possible start of construction in the absence of complete plans and specifications, for early-completion bonus, or for profit-sharing arrangements as incentives to the contractor.

One alternative often used is a cost-plus-fixed-fee contract. When this is used, the contractor is **reimbursed** for the cost plus a fixed amount, the fee for accomplishment of the work. After the scope of the work has been clearly defined and both parties have agreed on the estimated cost, the amount of the contractor's fee is determined in relation to character and volume of work involved and the duration of the project. Thereafter the fee remains fixed, regardless of any **fluctuation** in actual cost of the project. There is no incentive for the contractor to inflate the cost under this type of contract since the contractor's fee is unaffected thereby. But maximum motivation toward efficiency and quick completion inherent in fixed-price contracts may be lacking.

A profit-sharing clause issometimes written into the cost-plus-fixed-fee contracts as an incentive for the contractor to keep cost at a minimum, allowing the contractor a share of the savings if the actual cost, upon completion, underruns the estimated cost. This provision may also be accompanied by a penalty to be assessed against the contractor's fee in case the actual cost exceeds the agreed estimated cost.

A fundamental requirement for all cost-plus-fixed-fee contract agreements is a definition of cost. A clear distinction should be made between reimbursable costs and costs that make up the contractor's general expense, payable out of the contractor's fee. Some contracts, which would otherwise run smoothly, become difficult because of failure to define cost clearly. Usually, only the cost directly and solely assignable to the project is reimbursed to the contractor. Therefore, the contractor's central office overhead and general expense, salaries of principals and headquarters staff, and interest on capital attributable to the project frequently come out of the fee, although a fixed allowance in cost for contractor's home-office expense may be allowed.

Cost-plus-fixed-fee contracts do not guarantee a profit to the contractor. They may also result, particularly in government cost-plus-fixed-fee contracts, in unusually high on-job overhead occasioned by frequent government requirements for **onerous** and **cumbersome** procedures in accountability and accounting.

General contractors generally obtain subcontract and material-price bids before submitting a bid for a project to the owner. Usually, these bids are incorporated into the subcontracts. (Sometimes, general contractors continue shopping for subcontract bids after the award of the general contract, to attain budget goals that may have been exceeded by the initial bids.)

For every project, the contractor should keep records of everything to be purchased for the job and prepare a budget for each of the items. As each subcontract is awarded, the contractor should enter the subcontractor's name and the amount of the subcontract. Later, the profit or loss on the purchase should be entered in the record, thus maintaining a continuous **tabulation** of the status of the purchase. For convenience, priority numbers may be assigned to the various items, in order of preference in purchasing. Examination of the numbers enables a contractor to concentrate efforts on the subcontracts that must be awarded first.

Contractors typically **solicit** bids from subcontractors employed previously with satisfactory results and through notices in trade publications, such as The Dodge Bulletin. If the owner or the law requires use of specific categories of subcontractors, the contractor must obtain bids from qualified members of such categories. After receipt of subcontractor bids, the contractor should analyze and tabulate them for fair comparison. To make such a comparison, the contractor should ensure that the bidders for a trade are including the same items. For this purpose, the contractor should question each of the bidders, when necessary, and from the answers received tabulate the exact items that are included in or excluded from each bid. Although this may seem obvious, it should be reiterated that a good construction manager may alter the division of work among subcontractors to receive the most cost-efficient completion of work. If a subcontractor's proposal indicates that a portion of the work is being omitted, the contractor should cross-check the specifications and other trades to be purchased to determine if the missing items are the province of other subcontractors.

Various forms are available for use as subcontract agreements. The standard form, "Contractor-Subcontractor Agreement," A401, American Institute of Architects, is commonly used. A subcontract rider **tailored** for each job usually is desirable and should be initialed by both parties to the contract and attached to all copies of the subcontract. The rider should take into account modifications required to adapt the standard form to the job. It should cover such items as start and completion dates, options, alternatives, insurance and bonding requirements, and special requirements of the owner or leading agency.

To achieve a fair distribution of risks and provide protective techniques for the benefit of both parties, it is necessary for subcontracts to be carefully drawn. The prime contractor wishes to be assured that the subcontractor will perform in a timely and efficient manner. On the other hand, the subcontractor wishes to be assured of being promptly and fairly compensated and that no onerous burdens of performance or administration will be imposed.

Basic problems arise where parties fail to agree with respect at least to the essentials of the transaction, including the scope of work to be performed, price to be paid, and performance. The subcontract must include the **regulatory** requirements of the prime contract and appropriate arrangements for price, delivery, and specifications. It is insufficient to assume that writing a subcontractor a purchase order binds that subcontractor to the terms of the prime contractor's agreement. Subcontracts should be explicit with respect to observance of the prime contract. Also, subcontractors should be fully informed by being furnished with the prime-contract plans, specifications, and other construction documents necessary for a complete understanding of the obligations to which they are bound.

Although prime contracts often provide for approval of subcontractors as to fitness and responsibility, the making of a subcontract establishes only indirect relationships between owner and subcontractor. The basis upon which subcontract agreements are drawn on fixed-price work is of no concern to the owner because the prime contractor, by terms of the agreement with the owner, assumes complete responsibility. Under cost-plus prime contracts, however, subcontracts are items of reimbursable cost. As such, their terms, particularly the monetary considerations involved, are properly subject to the owner's approval.

Subcontract agreements customarily define the sequence in which the work is to be done. They also set time limits on the performance of the work. Nevertheless, prime contractors are reluctant to delegate by means of subcontracts portions of a project where failure to perform might have serious consequences on completion of the whole project—for example, the construction of a tunnel for diversion of water in dam construction.

In the heavy-construction industry, the greater the risk of loss from failure to perform, the less work is subcontracted. Such damages as may be recovered under subcontract agreements for lack of performance are usually small recompense for the overall losses arising out of the **detrimental** effect on related operations and upon execution of the construction project as a whole.

This situation has given rise to a common trade practice in the heavy-construction industry: The prime contractor builds up a following of subcontractors known for their ability to complete commitments properly and on time and generally to cooperate with and fit into the contractor's job operating team. The prime contractor often negotiates subcontracts or limits bidding to a few such firms. As a result, the same subcontractors may follow the prime contractor from job to job.

Retainage. Prime contracts require, as a rule, that a percentage (usually 10%) of the contractor's earnings be retained by the owner until final completion of the job and acceptance by the owner. Unless otherwise arranged, the provisions of the prime contract regarding payment and retainage pass into the subcontract. This is done with the usual stipulation in the prime contract that makes the subcontract subject to all the requirements of the prime contract.

Subcontractors whose work, such as site clearing, access-road building, or excavation, is performed in the early construction stages of a project may be severely impacted financially by this retainage. The standard retainage provisions may result in their having to wait a long time after completion of their work to collect the retained percentage. So the retainage on the general run of subcontracts, particularly those for work in the early phases of a project, often is reduced to a nominal amount after completion of the subcontractor's work. Justification for waiting until final completion of the job and acceptance by the owner may exist, however, under subcontracts for installed equipment carrying performance guarantees or for other items with vital characteristics.

An agreement may be negotiated, however, for early release or reduction of retainage. The subcontract should be specific in the matter of payment and release of retained earnings.

A contractor should never bid a job without first thoroughly examining the site. This should be done early enough for the owner to have sufficient time to issue **addenda** to the plans and specifications, if required, to clarify questionable items.

Before visiting the site, the contractor should prepare a checklist of items to be investigated. The checklist should include, where applicable, the following: transportation facilities, electric power supply, water supply, source of construction materials, type of material to be encountered in required excavation or borrow pits, possible property damage from blasting and other operations of the contractor, interference from traffic, available labor supply (number and length of shifts per week being worked in the **vicinity**), areas available for construction of special plant, location of waste-disposal areas and access thereto, and weather records if not otherwise available.

New Words and Expressions
[1] instrumentality [ɪnstrʊmen'tælɪtɪz] n. 工具；手段；
[2] reimburse [ˌriːɪm'bɜːs] n. 报销，偿还；
[3] fluctuation [ˌflʌktʃʊ'eɪʃn] n. 波动，涨落，起伏；
[4] onerous ['əʊnərəs] adj. 繁重的；
[5] cumbersome ['kʌmbəsəm] adj. 笨重的；累赘的；
[6] tabulation [ˌtæbjʊ'leɪʃn] n. 制表；
[7] solicit [sə'lɪsɪt] v. 恳请；
[8] tailored ['teɪləd] adj. 定制的；
[9] regulatory ['regjələtəri] adj. 监督的；管理的；
[10] detrimental [ˌdetrɪ'mentl] adj. 有害的；
[11] retainage n. 保留款；
[12] addenda [ə'dendə] n. 附录，补遗；
[13] vicinity [və'sɪnətɪ] v. 附近，邻近。

LESSON 13　BUILDING INFORMATION MODELING

Text A　Basic Concepts of BIM

Brief Introduction

Building Information Modeling (BIM) is one of the most promising developments in the architecture, engineering, and construction (AEC) industries. With BIM technology, one or more accurate **virtual** models of a building are constructed digitally. They support design through its phases, allowing better analysis and control than manual processes. When completed, these computer generated models contain precise geometry and data needed to support the construction, fabrication, and procurement activities through which the building is realized.

BIM also accommodates many of the functions needed to model the **lifecycle** of a building, providing the basis for new design and construction capabilities and changes in the roles and relationships among a project team. When adopted well, BIM facilitates a more integrated design and construction process that results in better quality buildings at lower cost and reduced project duration.

THE CURRENT AEC BUSINESS MODEL

Currently, the facility delivery process remains fragmented, and it depends on paper-based models of communication. Errors and omissions in paper documents often cause unanticipated field costs, delays, and eventual lawsuits between the various parties in a project team. Efforts to address such problems have included the use of real-time technology, such as project Web sites for sharing plans and documents; and the implementation of 3D CAD tools. Though these methods have improved the timely exchange of information, they have done little to reduce the severity and frequency of conflicts caused by paper documents or their **electronic equivalents.**

One of the most common problems associated with 2D-based communication during the design phase is the considerable time and expense required to generate critical assessment information about a proposed design, including cost estimates, energy-use analysis, structural details, and so forth. These analyses are normally done last, when it is already too late to make important changes. Because these iterative improvements do not happen during the design phase, value engineering must then be undertaken to address inconsistencies, which often results in compromises to the original design.

From CAD to BIM

All CAD systems generate digital files. Older CAD systems produce plotted drawings. They generate files that consist primarily of vectors, associated line-types, and layer identifications. As these systems were further developed, additional information was added to these files to allow for blocks of data and associated text. With the introduction of 3D modeling, advanced definition and complex surfacing tools were added.

As CAD systems became more intelligent and more users wanted to share data associated with a given design, the focus shifted from drawings and 3D images to the data itself. A building model produced by a BIM tool can support multiple different views of the data contained within a drawing set, including 2D and 3D. A building model can be described by its content or its capabilities. The latter approach is preferable, because it defines what you can do with the model rather than how the database is constructed.

BIM moves the industry forward from current paper-centric processes (3D CAD, animation, linked databases, and 2D CAD drawings) toward an integrated and interoperable workflow where these tasks are collapsed into a coordinated and collaborative process that maximizes computing capabilities, Web communication, and data aggregation into information and knowledge capture.

WHAT IS NOT BIM TECHNOLOGY?

The term BIM is a popular buzzword used by software developers to describe the capabilities that their products offer. As such, the definition of what constitutes BIM technology is subject to variation and confusion. To deal with this confusion, it is useful to describe modeling solutions that do not utilize BIM design technology. These include tools that create the following kinds of models:

Models that contain 3D data only and no (or few) object attributes. These are models that can only be used for graphic visualizations and have no intelligence at the object level. They are fine for visualization but provide little or no support for data integration and design analysis.

Models with no support of behavior. These are models that define objects but cannot adjust their positioning or proportions because they do not utilize parametric intelligence. This makes changes extremely labor intensive and provides no protection against creating inconsistent or inaccurate views of the model.

Models that are composed of multiple 2D CAD reference files that must be combined to define the building. It is impossible to ensure that the resulting 3D model will be feasible, consistent, countable.

Models that allow changes to dimensions in one view that are not automatically reflected in other views. This allows for errors in the model that are very difficult to detect.

WHAT ARE THE BENEFITS OF BIM

BIM technology can support and improve many business practices. Though it is unlikely that all of the advantages discussed below are currently in use, we have listed them to show the entire changes that can be expected as BIM technology develops.

Benefits to owner

Increased Building Performance and Quality

Developing a schematic model prior to generating a detailed building model allows for a more careful evaluation of the proposed scheme to determine whether it meets the building's functional and sustainable requirements. Early evaluation of design **alternatives** using analysis/simulation tools increases the overall quality of the building.

Improved Collaboration Using Integrated Project Delivery

When the owner uses Integrated Project Delivery (IPD) for project procurement, BIM can be used by the project team from the beginning of the design to improve their understanding of project requirements and to extract cost estimates as the design is developed. This allows design and cost to be better understood and also avoids the use of paper exchange and its associated delays.

Design Benefits

Earlier and More Accurate Visualizations of a Design

The 3D model generated by the BIM software is designed directly rather than being generated from multiple 2D views. It can be used to visualize the design at any stage of the process with the expectation that it will be dimensionally consistent in every view.

Automatic Corrections When Changes Are Made to Design

If the objects used in the design are controlled by parametric rules that ensure proper **alignment**, then the 3D model will be free of geometry and alignment errors. This reduces the users' work to manage design changes.

Generation of Accurate and Consistent 2D Drawings at Any Stage of the Design

Accurate and consistent drawings can be extracted for any set of objects or specified view of the project. This significantly reduces the amount of time and number of errors associated with generating construction drawings.

Earlier Collaboration of Multiple Design Disciplines

BIM technology facilitates simultaneous work by multiple design disciplines. While collaboration with drawings is also possible, it is more difficult and time consuming than working with one or more coordinated 3D models in which change control can be well managed. This shortens the design time and significantly reduces design errors and omissions.

Easy Verification of Consistency to the Design Intent

BIM provides earlier 3D visualizations and quantifies the area of spaces and other material quantities, allowing for earlier and more accurate cost estimates. For technical buildings (labs, hospitals, and the like), the design intent is often defined quantitatively, and this allows a building model to be used to check for these requirements.

Improvement of Energy Efficiency and Sustainability

Linking the building model to energy analysis tools allows evaluation of energy use during the early design phases. This is not practical using traditional 2D tools because of the time required to prepare the relevant input. If applied at all, energy analysis is performed at the end of the 2D design process as a check or a regulatory requirement, thus reducing the opportunities for modifications that could improve the building's energy performance.

Construction and Fabrication Benefits

Use of Design Model as Basis for Fabricated Components

If the design model is transferred to a BIM fabrication tool and detailed to the level of fabrication objects, it will contain an accurate representation of the building objects for fabrication and construction. Because components are already defined in 3D, their automated fabrication using numerical control machinery is facilitated. Such automation is standard practice today in steel fabrication and some sheet metal work.

Quick Reaction to Design Changes

The impact of a design change can be entered into the building model and changes to the other objects in the design will automatically update. Some updates will be made automatically based on the established parametric rules. Additional cross-system updates can be checked and updated visually or through clash detection. The consequences of a change can be accurately reflected in the model and all subsequent views of it. In addition, design changes can be resolved more quickly in a BIM system because modifications can be shared, estimated, and resolved without the use of time-consuming paper transactions.

Discovery of Design Errors and Omissions before Construction

Because the virtual 3D building model is the source for all 2D and 3D drawings, design errors caused by inconsistent 2D drawings are eliminated. In addition, because models from all disciplines can be brought together and compared, multisystem interfaces are easily checked both systematically and visually. Conflicts and constructability problems are identified before they are detected in the field. Coordination among participating designers and contractors is enhanced and errors of omission are significantly reduced. This speeds the construction process, reduces costs, minimizes the likelihood of legal disputes.

New Words and Expressions

[1] virtual ['vɜːtʃuəl] *adj.* 虚拟的；

[2] lifecycle [laɪfsaɪkl] *n.* 寿命周期；

[3] electronic equivalents *n.* 等效电子文档；

[4] alternative [ɔːl'tɜːnətɪv] *n.* 备选的事物；

[5] alignment [ə'laɪnmənt] *n.* 排成直线；

[6] fabrication [fæbrɪ'keɪʃn] *n.* 制造。

参考译文：建筑信息模型（BIM）基本知识

简介

建筑信息模型（Building Information Modeling，BIM）是建筑、工程和建造［architecture，engineering，and construction（AEC）］行业最具发展前景的技术之一。通过BIM技术可以进行一个或多个建筑数字化虚拟建造。它可以支持各阶段的设计，比人工设计方法更利于分析和控制。当模型完成后，计算机可以生成具有精确几何尺寸的模型，生成的数据可以用于辅助建设、制作、采购等。

BIM 也为模拟建筑的全寿命周期性能提供了方法，为新的设计、建造方式以及在一个项目团队中改变角色和关系提供了基础。当使用得当时，BIM 能实现更紧密的设计和建造过程，并提升建筑质量、减小成本和缩短项目时间。

现有的 AEC 商业模型

现有的设备提供构成仍比较零散，仍用纸面化的模型进行沟通。纸质模型的错误常常导致预料之外的现场成本和延误，会导致工程各方的法律诉讼。为解决这些问题已经采取了工程网站等实时技术来分享计划、文件和 3D 的 CAD 工具。尽管这些方法已经提高了信息交换的及时性，但对于减小纸质文件或同类的电子文件导致的频繁的冲突作用甚小。

在设计阶段采用 2D 模型进行沟通所产生的最常见的问题之一是得到一种设计方案的关键评判信息（如成本预估、能源预估、结构细节等）需要消耗大量时间和金钱。这些分析通常在最后才做，此时再做重要修改已经太迟了。因为这些交互改进并没有发生在设计阶段，之后只能采用价值工程处理不一致，这使得最终结果常常是对最初设计进行妥协。

从 CAD 到 BIM

所有的 CAD 系统都产生数字文件。旧的 CAD 系统绘制图形。这些文件包含向量以及线型、图层信息。随着对这些系统的进一步开发，这些文件又增加了块和文字信息。随着 3D 建模的引入，高级的面层定义和处理工具也加入了进来。

随着 CAD 系统变得更智能，更多的使用者想要分享设计信息而不是分享 3D 图形。BIM 工具可以提供一个图形集的不同信息，包括 2D 和 3D 的。建筑模型可以按照它包含的内容来描述，或者按照它的功能来描述。更喜欢使用后者，因为后者定义了可以用这个模型做什么而不是描述这个数据库是怎么建的。

BIM 技术将现有的以纸质文件为中心的流程（3D CAD 模拟、数据库连接以及 2D CAD 图形）转变为统一的能共同协作的工作流程。在新工作流程中，任务被打碎为协调的协作过程，并将计算能力最大化；网络沟通和信息聚集转变为信息和知识的获取。

什么不是 BIM 技术

BIM 这个术语是软件编写者用于描述产品功能的。因此，哪些构成了 BIM 技术有很多分歧和混淆。避免混淆需要描述下哪些不属于 BIM 技术。这包括能生成这些模型的工具：

只包含 3D 信息，不包含其他属性的模型。这些模型只能用于图像展示，并不具有实物层面的信息。它们对于图像化有用，但对于信息集成和设计分析没有作用。

没有行为的模型。这些模型定义了实物，但该实物不具有参数，因此其位置或配比不能调整。这使得对该物体的修改非常耗费人工，而且不能防范模型不一致或不准确。

由多个 2D CAD 文件构成的模型，必须合在一起才能定义建筑。这个不能保证生成的 3D 模型是有效、一致且可靠的。

允许对模型的某一维度修改，但修改并不自动反映到其他视图上。这会产生很多难以察觉的错误。

BIM 的好处

BIM 技术可以支持和改进很多商业经营。尽管下面的优点在现有应用中并未全部实现，我们还是把 BIM 所能带来的所有改变全都列举出来。

对业主的好处

提高建筑表现和建筑质量

在建立细致的建筑模型前先建一个概略模型，可以准确地评判它是否满足建筑的功能和耐久性要求。用分析和模拟工具对参选方案进行早期评判可以改进建筑的整体质量。

用项目集成交付模式提升合作水平

当业主采用建筑集成交付模式进行采购时，项目团队采用BIM技术可以在项目初期就提升对项目要求的理解，随着设计深入可以更好地估计成本。这使得设计和成本能更好地被理解，并避免纸质文件交互带来的延误。

设计方的好处

对设计更早更准确的可视化

BIM软件生成的3D模型可直接展示设计，而不是通过各个2D视角来呈现模型。这在设计的各个阶段都可以用于对设计的可视化，保证模型从各个角度看都是一致的。

当设计变更时可以自动做修正

如果物体由参数控制，则3D模型不会出现几何或一致性的错误。这能降低设计变更时的工作量。

在设计的每个阶段都能生成准确的2D图形

可以从实物集合或项目的特定视角提取准确一致的图形。这可以显著减少制作施工图所需的时间和所产生的错误。

多个设计专业可以进行早期合作

BIM技术使得多个设计专业可以同时工作。通过图形进行协作也可以，但这相对于用3D模型协作更难、更耗时，在3D模型中很好地管理修改。这缩短了设计时间，减少了设计的错误和遗漏。

更容易检查是否满足设计目的

BIM技术在初期就可以提供3D模型，将面积、空间及材料量化，使得可以更准确地估计造价。对于专业建筑（实验室、医院等），设计目的往往需要量化，BIM使得建筑模型可以用于查看是否满足这些设计要求。

提升能源效率和可持续性

将建筑模型与能量分析相关联可以在设计初期分析能量消耗。这在使用传统的2D工具时是实现不了的，因为没有时间提供所需的输入。如果在2D设计的末期进行能量分析，可以查看是否满足通常的要求。这减少了修改建筑方案提升其能源利用效率的机会。

建造方的好处

将设计模型作为构件制作的基准

如果设计模型导入了BIM制作工具并达到了制作等级的细致程度，这将能准确反映所需制作的建筑构件。由于定义了3D构件，可以使用数控机床自动制作这些构件。这种自动制作方式如今在钢结构生产中已成标准生产工艺。

对设计变更可以快速响应

设计变更可以导入建筑模型并且其他物体会根据变更自动更新。一些更新是根据建立的参数规则自动实现的。其他跨系统的更新需要检查一下，并进行碰撞检查。一项变更的后果可以在模型以及模型的各个视角中准确地反映出来。此外，设计变更可以在BIM系统中更快地解决，因为修正可以直接分享、评判和处理，不需要花时间制作纸质文件。

在建造前发现设计错误和遗漏

因为虚拟的 3D 建筑模型是所有 2D 和 3D 图形的来源，2D 模型中的不一致错误就可以避免了。此外，因为所有设计专业的模型可以一起进行比较，多系统的界面可以很容易被系统化、视觉化地比较。在进场施工前就可以发现冲突和建造问题。设计者和承包方的合作可以加强，错误和遗漏可以显著减少。这加速了建设过程，减少了成本和法律争端的可能性。

Text B PARAMETRIC MODELING OF BUILDINGS

In manufacturing, parametric modeling has been used by companies to embed design, engineering, and manufacturing rules within the parametric models of their products. Using parametric modeling, companies usually define how their object families are to be designed and structured, how they can be varied parametrically and related into **assemblies** based on function, production, assembly, and other **criteria**. The companies are embedding corporate knowledge based on past manual efforts on design, production, assembly, and maintenance concerning what works and what does not. This is the standard practice in large aerospace, manufacturing, and electronics companies.

Parametric Design

Conceptually, building information modeling tools are different flavors of object-based parametric modeling systems. They are different because they have their own predefined set of object classes, each having possibly different behaviors programmed within them, as outlined above.

In addition to **vendor**-provided object families, a number of Web sites make additional object families available for downloading and use. These are the modern equivalent of drafting block libraries that were available for 2D drafting systems (but, of course, they are much more useful and powerful). They include, for example, furniture, plumbing and electrical equipment, and proprietary fasteners for concrete fabrication. They are available both as generic objects and as models of specific products.

The built-in behaviors of BIM objects identify how they can be linked into assemblies and automatically adjust their own design when their context with other objects change. Examples are walls and their updates when other walls or ceilings change. Another is how spaces update in most systems when their bounding walls change. These object classes also define what features can be associated with building objects.

A functional difference in building modeling tools from that of other industries is the need to explicitly represent the space enclosed by building elements. Environmentally conditioned building space is a primary function of a building. The shape, volume, surfaces, environmental quality, lighting, and other properties of an interior space are critical aspects to be represented and assessed in a design.

Until recently architectural CAD systems were not able to represent building spaces explicitly; objects were approximated using a drafting system approach, as user-defined polygons with an associated space name. Credit is due to the General Services Administration (GSA) for demanding that BIM design applications be capable of automatically deriving and **updating** space volumes, beginning in 2007. Today, most BIM design applications represent a building space as an automatically generated and updated polygon defined by the wall intersections with a floor slab. The polygon is then extruded to the average ceiling height or possibly trimmed to a sloping ceiling surface. The older manual method has all the weaknesses of manual drafting: users must manage the consistency between wall boundaries and spaces, making updates both tedious and error-prone. The new definition is not perfect: it works for vertical walls and flat floors, but ignores vertical changes in wall surfaces, and often cannot reflect nonhorizontal ceilings.

Architects work initially with nominal building element shapes. But engineers and fabricators must deal with fabricated shapes and layouts that vary from nominal and must carry fabrication-level information. Also, shapes change due to pre-tensioning (camber and foreshortening), deflect due to gravity, and expand and contract with temperature. As building models become more widely used for direct fabrication, these aspects of parametric model shape generation and editing will require additional capabilities of BIM design applications.

Parametric modeling is a critical productivity capability, allowing low-level changes to update automatically. 3D modeling would not be productive in building design and production without the automatic update features made possible by parametric capabilities. However, there are hidden effects. Each BIM tool varies with regard to the level of implementation of parametric modeling, the parametric object families it provides, the rules embedded within it, and the resulting design behavior. Customizing the behaviors of the object classes provided involves a level of new expertise not widely available in current architecture, engineering, and fabrication offices.

Parametric Modeling for Construction

While BIM design intent applications allow users to assign layers to a wall section in terms of a 2D section, some architectural BIM design applications include parametric layout of nested assemblies of objects, such as stud framing, within a layer of a generic wall. This allows generation of the detailed framing and derivation of a cut lumber schedule, reducing waste and allowing for faster erection of wood or metal stud-framed structures. In large-scale structures, similar framing and structural layout options are necessary operations for fabrication. In these cases, objects are parts which are composed into a system (structural, electrical, piping, and the like) and the rules determine how the components are organized. **Components** often have features, such as connections, that are custom designed and fabricated. In the more complex cases, each of the system's parts are then internally composed of their constituent parts, such as steel reinforcing in concrete or complex framing of long-span steel structures.

A distinct set of BIM design applications have been developed for modeling at the more detailed fabrication levels. These tools provide different object families for embedding different types of expertise. They are also related to different specific uses, such as materials tracking and ordering, plant management systems, and automated fabrication software.

Recent advances have been made in concrete engineering with cast-inplace and precast concrete. The layout can be easily adjusted to the section size and to the layout of columns and beams. Parametric modeling operations can include shape subtraction and addition operations that create reveals, notches, bullnoses, and cutouts defined for connections to other parts.

In fabrication modeling, detailers refine their parametric objects for well understood reasons: to minimize labor, to achieve a particular visual appearance, to reduce the mixing of different types of work crews, or to minimize the types or sizes of materials. Standard design-guide implementations typically address one of multiple acceptable approaches for detailing. In some cases, various objectives can be realized using standard detailing practices. In other circumstances, these detailing practices need to be overridden. A company's best practices or standard interfacing for a piece of fabrication equipment may require further customization. In future decades, design handbooks will be supplemented in this way, as a set of parametric models and rules.

A critical difference between these earlier systems and BIM is that users can define much more complex structures of object families and relations among them than is possible with 3D CAD, without undertaking programming-level software development. With BIM, a curtain wall system attached to columns and floor slabs can be defined from scratch by a knowledgeable nonprogrammer. Such an endeavor would require the development of a major application extension in 3D CAD.

User-Defined Parametric Objects

Each BIM design application has an expanding set of predefined parametric object classes, reflecting its target functionality. The architectural BIM applications' predefined objects generally capture conventions of design intent for architectural design. Currently, these also are frequently used to capture construction management (CM) information for construction coordination. However, the objects used in CM require additional information, dealing with tasks and schedules, material tracking, and other management links. Other applications have been developed, with different objects, for representing structural design and analysis information, and still others for representing information for different building subsystems, such as mechanical systems, plumbing, or electrical systems. Some applications focus on the design-intent level of detail and others at the fabrication level.

Each BIM application and the predefined objects that come with it are meant to capture the standard conventions in the area of building that the application targets. Most design and engineering domains have handbooks of standard practice. In architecture, this has for a long period been addressed by Ramsey and Sleeper's Architectural Graphic Standards (Ramsey and Sleeper 2000). In other areas, standard practice is captured by handbooks such as the AISC handbook Detailing for Steel Construction (AISC 2007), or the PCI Design Handbook (PCI 2004). Standard practice reflects industry conventions, how to design building parts and systems, based on current practices, often addressing safety, structural performance, material properties, and usage. Design behavior, on the other hand, has not been codified, resulting in different object behaviors in each of the BIM design tools. The base objects in each different BIM design tool is a repackaging of standard practice, as interpreted by the software company's software developers, often with input from industry groups and experts.

In the real world, however, these predefined objects and their built-in behaviors will be limiting at the design and fabrication stages, for a variety of reasons, some enumerated below:

- A different configuration of parts is desired for construction, analysis, or aesthetic reasons.
- The base parts do not address a specific design condition encountered in a design or real-world context.
- A building system whose structure and behavior is not available by the software or building system vendors.
- Some objects are not provided by the BIM design application.
- Improved objects incorporating company best practices.

If a needed parametric object capability does not exist in the BIM tool, the design and engineering team has these options:

1. Creating an object in another system and importing it into your BIM tool as a reference object, without local editing capabilities.

2. Laying out the object instance manually using solid modeling geometry, assigning attributes manually, and rememberingto update the object details manually as needed.

3. Defining a new parametric object family that incorporates the appropriate external parameters and design rules to support automatic updating behaviors, but the updates are not related to other object classes.

4. Defining an extension to an existing parametric object family that has modified shape, behavior, and parameters; the resulting object (s) fully integrate with the existing base and extended objects.

5. Defining a new object class that fully integrates and responds to its context.

The first two methods listed above reduce the capabilities of piece editing to the CAD-level, without parametric representation. All BIM model generation tools support the definition of custom object families (points 3 and/or4). These allow users to define new object classes that can update according to the context defined within them. More challenging is the integration of new custom objects with existing predefined objects such as doors, walls, slabs, and roofs that are provided by the BIM tool. New objects need to fi t into the BIM platform's already-defined updating structures; otherwise, the interfaces of these objects with others must be edited manually. If a firm frequently works with some building type or system involving special object families, the added labor to define these parametrically is easily justified. They provide automatic application of company best practices in the various contexts found in different projects and can be applied firmwide. These may be at a high level for layouts or those needed for detailing and fabrication.

New Words and Expressions
[1] assembly [ə'sembli] n. 聚集；
[2] criteria [kraɪ'tɪərɪə] n. 评判准则；
[3] vendor ['vendə(r)] n. 卖主；
[4] update ['ʌpdeɪt] n. 更新；
[5] component [kəm'pəʊnənt] n. 构件。

附录 土木工程专业词汇汉英表达

附录 A 结构工程与防灾工程

area of section 截面面积
axial pressure 轴向压力
bar splicing 钢筋搭接
beam 梁
bearing capacity 承载力
bearing structure 承重结构
bending failure 弯曲破坏
bending strength 抗弯强度
bent-up bar 弯起钢筋
brittle failure 脆性破坏
brittleness 脆性
cantilever beam 悬臂梁
cast-in-place reinforced concrete 现浇钢筋混凝土结构
cement 水泥
cohesive force 粘结力
column 柱
compressive region 受压区
compressive strength 抗压强度
concentrated load 集中荷载
concentration of stresses 应力集中
concrete 混凝土
concreting 浇注混凝土
construction 构造
cover to reinforcement 钢筋保护层
crack 裂缝
crane beam 吊车梁
cranked slab stairs 板式楼梯
cross section 横截面
deflection 挠度

design load 设计荷载
design strength 设计强度
displacement 位移
distribution load 分布荷载
distribution steel 分布钢筋
ductility 延性
eccentric compression 偏心受压
eccentric distance 偏心距
eccentric load 偏心荷载
eccentric tension 偏心受拉
elastic modulus 弹性模量
external force 外力
fatigue strength 疲劳强度
flat slab 无梁楼盖
foundation 基础
frame structure 框架结构
hoop reinforcement 箍筋
limit of yielding 屈服极限
load 荷载
loading 加载
loss of prestress 预应力损失
lower limit of yield 屈服强度下限
main beam 主梁
member in bending 受弯构件
midspan load 跨中荷载
moment 弯矩
non-reinforced concrete 素混凝土
percentage of elongation 延伸率
plasticity 塑性
Poisson's ratio 泊松比

pouring 浇注
precast slab 预制板
pressure 压力
prestressed reinforced concrete 预应力钢筋
混凝土
prestressed reinforcement 预应力钢筋
reinforced concrete structure 钢筋混凝土结构
reinforced concrete（RC）钢筋混凝土
reinforcement ratio 配筋率
reinforcing steel bar 钢筋
reliability 可靠性
roof truss 屋架
secondary beam 次梁
shear deformation 剪切变形
shear modulus 剪切模量
shear wall 剪力墙
shear 剪力
simple beam 简支梁
slab 板
span 跨度

span-to-depth ratio 跨高比
state of stress 应力状态
steel wire 钢丝
stiffness，rigidity 刚度
stirrup ratio 配箍率
stirrup spacing 箍筋间距
strain 应变
stress diagram 应力图
stress relaxation 应力松弛
stress 应力
stress-strain curve 应力应变曲线
tensile region 受拉区
tensile strength 抗拉强度
tension 拉力
torsional strength 抗扭强度
two-way reinforcement 双向配筋
yield load 屈服荷载
yield point 屈服点
yield strength 屈服强度
yield 屈服

附录 B 桥 梁 工 程

abutment 桥台
accidental combination for action effects
  作用效应偶然组合
allowable stress | permissible stress 容许应力
allowable stresses method 容许应力法
allowable value of crack width 裂缝宽度容
  许值
anchor eye 锚孔
approach ramp 引道坡
approach road 引道
arch springing hinge 拱脚铰
arch springing 拱脚
arch without articulation 无铰拱
area of infection 受影响面
auxiliary bridge 便桥

average annual flood 平均年洪水流量
axial force 轴力
axial tension 轴向拉力，轴向拉伸
balanced steel ratio 平均配筋率
bank erosion 河岸侵蚀
bar system 杆系
batter pile 斜桩
beam soffit 梁拱腹
bearing platform 承台
bearing stiffener 支承加劲肋
bending stiffness 弯曲刚度
bent cap 盖梁
box girder 箱梁
breaking strength 破坏强度
bridge aseismatic strengthening 桥梁抗震加固

bridge best practice guidelines 桥梁最佳施工指南
bridge clearance 高潮时桥梁净空高度
bridge vibration 桥梁振动
brittle failure 脆性破坏
buried abutment 埋置式桥台
cable-stayed bridge 斜拉桥
calculated span 计算跨径
cantilever 悬臂梁
car load 车辆荷载
cast-in-place pile 灌注桩
catenary 悬链线
center of moments 弯矩中心
central divider 道路中央分隔栏
characteristic value of an action 作用标准值
characteristic value of permanent action 永久作用标准值
characteristic value of strength of steel bar 钢筋强度标准值
characteristic value of strength of steel 钢材强度标准值
characteristic value of variable action 可变作用标准值
clamped supported beam 固定-简支梁
clear span 净跨距
coefficient of rigidity 刚性系数
coefficient of stabilization 稳定系数
combination for action effects 作用效应组合
combined dead load 组合恒载
complete shear failure 纯剪切破坏
compressive stress 压应力
concentrated load 集中荷载
concrete slab 混凝土板
construction documents design phase 施工图设计阶段
construction documents design 施工图设计
continuous beam on many supports 多跨连续梁
continuous beam 连续梁

coordinated geometric feature 配位几何特征
coping stone 墩台石
crack control 裂缝控制
crack 裂缝
crack-control reinforcement 控制裂缝钢筋
creep 徐变
crossing point 跨越点
cross-over pole 跨越杆
curvature 曲率
curved beam 曲梁
dead load 恒载
deflection curve 挠度曲线
deflection 挠度
design speed 设计速度
design value of a load 荷载设计值
design value of an action 作用设计值
designed flood frequency 设计洪水频率
direction of the maximum stress 最大应力方向
double lane 双车道
drag coefficient 阻力系数
drain pipe 泄水管
drainage system 排水系统
durability 耐久性
dynamic shear modulus of soils 动剪切模量
dynamic stiffness ratio 动刚度比
effective angle of internal friction 有效内摩擦角
elastic-plastic behavior 弹塑性状态
elastomeric bearing 弹性支承
emergency parking strip 紧急停车带
enhancement coefficient 增大系数
equation of the influence line 影响线方程
equivalent load 等效荷载
equivalent uniform live load 等效均布荷载
expansion joint 伸缩缝
fabricated bridge 装配式桥
failure limit state 破坏极限状态
fatigue fracture mechanics 疲劳断裂力学

fatigue strength 疲劳强度
final bending moment diagram 最终弯矩图
fixed arch 固端拱
floor live load 楼面活载
fracture mechanics 断裂力学
freely supported structure 简支结构
geometrical characteristic 几何特征
girder section 主梁截面
girder 主梁
graphical method for constructing shear and moment diagrams 绘制剪力和弯矩图的图解
gravity standard 重力标准值
headroom 净空高度
highway classification 公路等级
hollow slab beam 空心板梁
hooping 箍筋
hump bridge 拱桥
impact toughness 冲击韧性
influence line for reaction 反力影响线
influence line 影响线
influence surface 影响面
internal pressure stress 内压应力
intrados 拱腹线
inverse finite element analysis 反有限元分析
least favorable distribution 最不利分布
limit load design 极限荷载设计
limit state 极限状态
linear finite element method 线性有限元法
linkage parameter 杆系参数
list structure form 表结构形式
live load 活载
load bearing structure 荷载支承结构
load factor 荷载系数
load-factor method 荷载系数法
loading intensity 荷载强度
local shear failure 局部剪切破坏
longitudinal steel ratio 纵向配筋率
madeline 塔堡
main column（tower）主塔

main reinforcement 主钢筋
main spar 主梁翼
margin of safety 安全度
marginal strip 路缘带
maximum bending moment diagram 最大弯矩图
maximum shear strain 最大剪应变
maximum shear stress 最大剪应力
minimum steel ratio 最小配筋率
multi-cell box girder 多孔箱形梁
multi-layer 多层式
natural frequency of vibration 自振频率
nonlinear mechanics of plate and shell 板壳非线性力学
nonlinear random vibration 非线性随机振动
nonlinear vibration 非线性振动
non-uniform stress 非均布应力
normal point load 法向集中荷载
optimal structure designing 结构优化设计
outer support 外支座
parabolic cable 抛物线型钢丝束
partial safety factor 分项系数
pier body 墩身
pier cap 墩帽
pile cap 桩承台
pipe support 管支座
plain concrete 素混凝土
plane cross-section assumption 平截面假定
plate mechanics 板壳力学
plate unit 板单元
pontoon bridge 浮桥
precast concrete segmental ring 装配式预制混凝土环
preliminary dimension 初步尺寸
prestressed concrete 预应力混凝土
prestressed element 预应力元件
prestressed reinforcement 预应力钢筋
pre-stressing tendon 预应力钢筋束

pre-tensioned prestressing 先张法预应力
pre-tensioning 先张法
principal tensile stress 主拉应力
railway bridge engineering 铁路桥梁工程
railway engineering 铁路工程
rapid hardening cement 快硬水泥
ratio of rigidity 刚度比
reduction factor on number of waves 波数折减系数
reduction factor 折减系数
reinforcement ratio 配筋率
reliability 可靠度
representative valueof an action 作用代表值
representative value 代表值
residual stress 残余应力
rigid frame bridge 刚构桥
rigid frame 刚性构架
rigid joint 刚性接缝
rigid-frame beam 刚架梁
rise-span ratio 矢跨比
river pier 河墩
road capacity 道路容车量
road deck 道路面层
road divider 道路分隔栏
road works 道路工程
safety factor 安全系数
safety load factor 安全荷载系数
semi plastic stage 半塑性状态
serviceability limit state 正常使用极限状态
settlement joint 沉降缝
settlement 沉降
shear and moment diagrams 剪力和弯矩图
shear diagram 剪力图
shear failure 剪切破坏
shear nails 剪力钉
shear wall structure 剪力墙结构
shear wall 剪力墙
shearing force 剪力
sheet pile 板桩

shelving 排架
shoulder 路肩
side span & middle spin 边、中跨径
simple supported girder bridge 简支梁桥
single pier 单墩
skew bridge 斜交桥
soil stabilization works 土工加固工程
space truss structure 网架结构
span 跨度
specific creep 比徐变
splice bar 拼接板
standard highway vehicle load 公路车辆荷载标准
standard of structural stability 结构稳定性标准
standard value 标准值
static live load 静活载
statically determinate beam 静定梁
statically determinate rigid frame 静定刚架
steel bridge 钢桥
stiff girder connection 加劲梁节点
stiffening girder 加劲梁
stiffening rib 加劲肋
stiffness 刚度
strength reduction factor 强度折减系数
stress amplitude 应力幅值
stress analysis 应力分析
stress concentration 应力集中
stress direction 应力方向
stress intensification factor 应力增大系数
stress strain relation 应力应变关系
structural analysis 结构分析
structural control 结构控制
structural fortification 结构加固工程
structural safety 结构安全度
structural stability analysis 结构稳定性分析
structural stability 结构稳定性
structure of bar system 杆系结构
suboptimization 次最优化

support system 支撑体系
supporting member 支承构件
suspended deck structure 承台结构
suspension bridge 悬索桥
technology standard 技术标准
thermal stress analysis 热应力分析
three hinged arch 三铰拱
torsional shear strength 扭转剪切强度
total stress analysis 总应力分析
traffic capacity 容车量
traffic lane 行车道
traffic volume 车流量
truck loading 载重汽车荷载
trussed bridge 桁架桥
tunnel 隧道
two hinged arch 双铰拱
ultimate limit states 承载能力极限状态
ultimate load 极限荷载
uniform beam 等截面梁
uniform stress 均布应力
uniformly distributed load 均布荷载
unreliability 不可靠度
variable concentrated load 可变集中荷载
verification of serviceability limit states 正常使用极限状态验证
viaduct 高架桥
water level 水位
weather anchor 抗风锚
wind resistance coefficient 风阻力系数
wind resistance 抗风性
yield limit 屈服极限法

附录 C 岩土工程

active earth pressure 主动土压力
additional stress 附加应力
allowable amplitude of foundation 基础振动容许振幅
allowable bearing capacity of foundation soil 地基容许承载力
analysis of beams and slabs on elastic foundation 弹性地基梁（板）分析
anchor rod retaining wall 锚杆式挡土墙
anchored plate retaining wall 锚定板挡土墙
anchored sheet pile wall 锚定板板桩墙
angle of internalfriction 内摩擦角
anti-slide pile 抗滑桩
artificial foundation 人工地基
atterberg limits test 界限含水量试验
bearing capacity of foundation soil 地基承载力
bearing capacity of single pile 单桩承载力
bearing stratum 持力层
Biot's consolidation theory 比奥固结理论
Bishop method 毕肖普法
bore hole columnar section 钻孔柱状图
bored pile 钻孔桩
bottom heave＝basal heave （基坑）底隆起
bound water 结合水
boundary surface model 边界面模型
braced cuts 支撑围护
braced excavation 支撑开挖
braced sheeting 支撑挡板
bracing of foundation pit 基坑围护
caisson foundation 沉箱基础
califonia bearing ratio test 承载比试验
Cambridge model 剑桥模型
cantilever retaining wall 悬臂式挡土墙
cantilever sheet pile wall 悬臂式板桩墙
cap model 盖帽模型
capillary water 毛细管水
cast-in-place pile 灌注桩

centrifugal model test 离心模型试验
clay 黏土
coarse-grained soil（gravelly and sandy）粗粒土
coarse sand 粗砂
cobble 卵石
coefcent of compressibility 压缩系数
coefficient of active earth pressure 主动土压力系数
coefficient of consolidation 固结系数
coefficient of curvature 曲率系数
coefficient of earth pressur at rest 静止土压力系数
coefficient of gradation 级配系数
coefficient of passive earth pressure 被动土压力系数
coefficient of permeability 渗透系数
cohesion 黏聚力
compacting pile 夯实桩
compaction test 击实试验
compactness 密实度
compensated foundation 补偿性基础
composite foundation 复合地基
compression index 压缩指数
concentrated load 集中荷载
consistency 稠度
consoidated anisotropically undrained test 各向不等压固结不排水试验
consolidated anisotropically drained test 各向不等压固结排水试验
consolidated drained direct shear test 慢剪试验
consolidated drained triaxial compressure test 压密排水三轴压缩试验
consolidated drained triaxial test 固结排水试验
consolidated quick direct shear test 固结快剪试验
consolidated undrained test 固结不排水试验

consolidated undrained triaxial compression test 压密不排水三轴压缩试验
consolidation curve 固结曲线
consolidation pressure 固结压力
consolidation test 固结试验
consolidation under K_0 condition K_0 固结
constant head permeability test 常水头渗透试验
constitutive model 本构模型
contact pressure 接触压力
Coulomb's earth pressure theory 库仑土压力理论
counterforted retaining wall 扶壁式挡土墙
cpmpression curve 压缩曲线
creep 蠕变
critical edge pressure 临塑荷载
critical hydraulic gradient 临界水力梯度
critical state elastoplastic model 临界状态弹塑性模型
critical state soil mechanics 临界状态土力学
cross anisotropic model 层向各向同性体模型
cross strip footing 交叉条形基础
cushion 垫层法
cushion 换填法
cyclic loading 周期荷载
damping ratio 阻尼比
Darcy's law 达西定律
dead load＝sustained load 恒载，持续荷载
deep foundation 深基础
deep settlement measurement 深层沉降观测
deep well point 深井点
deformation 变形
degree of consolidation 固结度
degree of saturation 饱和度
density 密度
dewatering method 降低地下水位法
dewatering 基坑降水
dilatation 剪胀
direct shear apparatus 直剪仪

direct shear test 直剪试验
direct simple shear test 直接单剪试验
dispersion of waves 波的弥散特性
diving casting cast-in-place pile 沉管灌注桩
dry density 干密度
dry unit weight 干重度
Duncan-Chang model 邓肯－张模型
dynamic compaction replacement 强夯置换法
dynamic compaction 强夯法
dynamic magnification factor 动力放大因素
dynamic penetration test 动力触探试验
dynamic properties of soils 土的动力性质
dynamic settlement 振陷
dynamic shear modulus of soils 动剪切模量
dynamic simple shear test 振动单剪试验
dynamic strength of soils 动强度
dynamic subgrade reaction method 动基床反力法
dynamic triaxial test 动三轴试验
earth pressure at rest 静止土压力
earth pressure 土压力
earthquake engineering 地震工程
effective angle of internal friction 有效内摩擦角
effective cohesion intercept 有效黏聚力
effective stress approach of shear strength 剪胀抗剪强度有效应力法
effective stress failure envelop 有效应力破坏包线
effective stress strength parameter 有效应力强度参数
effective stress 有效应力
effective unit weight 有效重度
efficiency factor of pile groups 群桩效率系数
efficiency of pile groups 群桩效应
elastic deformation 弹性变形
elastic half-space foundation model 弹性半空间地基模型
elastic half-space theory of foundation vibration 基础振动弹性半空间理论
elastic model 弹性模型
elastic modulus 弹性模量
elastoplastic model 弹塑性模型
electro-osmotic drainage 电渗法
embedded depth of foundation 基础埋置深度
end-bearing pile 端承桩
engineering geological prospecting 工程地质勘察
equivalent lumped parameter method 等应变速率固结试验
evaluation of liquefaction potential 液化势评价
excavation 开挖（挖方）
excess pore water pressure 超孔压力
extended von Mises yield criterion 广义冯·米赛斯屈服准则
failure criterion 破坏准则
failure of foundation 基坑失稳
falling head permeability test 变水头渗透试验
fast consolidation test 快速固结试验
field permeability test 现场渗透试验
field vane shear strength 十字板抗剪强度
fill (soil) 填土
final set 最后贯入度
final settlement 最终沉降
fine sand 细砂
fine-grained soil (silty and clayey) 细粒土
flexible foundation 柔性基础
flow net 流网
flowing soil 流土
foundation design 基础设计
foundation engineering 基础工程
foundation vibration 基础振动
foundation wall 基础墙
free vibration column test 自振柱试验
free (resonance) vibration column test 自（共）振柱试验
freezing and heating 冷热处理法

friction pile 摩擦桩
frozen foundation improvement 冻土地基处理
frozen soil 冻土
general shear failure 整体剪切破化
geometric damping 几何阻尼
geostatic stress 自重应力
geotechnical engineering 岩土工程
geotechnical model test 土工模型试验
gradation 级配
grain size analysis test 颗粒分析试验
gravel 砂石
gravelly sand 砾砂
gravity retaining wall 重力式挡土墙
groundwater level 地下水位
group action 群桩作用
heave of base 基底隆起
high pressure consolidation test 高压固结试验
high-rise pile cap 高桩承台
homogeneous 均质
hydraulic gradient 水力梯度
hyperbolic model 双曲线模型
hyperelastic model 超弹性模型
ideal elastoplastic model 理想弹塑性模型
incremental elastoplastic theory 弹塑性增量理论
individual footing 单独基础
initial liquefaction 初始液化
in-situ pore water pressure measurement 原位孔隙水压量测
instability (failure) of foundation pit 基坑失稳
inverted beam method 倒梁法
isotropic 各向同性
K_0 consolidated drained test K_0固结排水试验
K_0 consolidated undrained test K_0固结不排水试验
landslide 滑坡
leakage 渗流
lime-soil compaction pile 灰土挤密桩

limit equilibrium method of slope 土坡稳定极限分析法
limit equilibrium method 极限平衡法
liquefaction strength 抗液化强度
liquefaction 液化
liquid limit test 液限试验
liquid limit 液限
liquidity index 液性指数
live load 活载
local shear failure 局部剪切破坏
logrithm of time fitting method 时间对数拟合法
major (intermediate, minor) principal stress 最大（中、最小）主应力
mat (raft) foundation 片筏基础
material damping 材料阻尼
maximum acceleration of earthquake 地震最大加速度
maximum dry density 最大干密度
mean diameter 平均粒径
medium sand 中砂
modified cambridge model 修正剑桥模型
modulus of compressibility 压缩模量
modulus of deformation 变形模量
Mohr-Coulomb failure condition 摩尔-库仑破坏条件
Mohr-Coulomb yield criterion 摩尔-库仑屈服准则
Mohr's envelope 摩尔包线
moist unit weight 湿重度
multi-dimensional consolidation 多维固结
natural foundation 天然地基
natural frequency of foundation 基础自振频率
natural period of soilsite 地基固有周期
negative skin friction of pile 负摩擦力
net foundation pressure 基底附加应力
nonlinear elastic model 非线性弹性模型
normal stresses 正应力
normally consolidated soil 正常固结土

numerical geomechanics 数值岩土力学
one dimensional consolidation 一维固结
open-end caisson foundation 沉井基础
optimum water content 最优含水量
over consolidation ration（OCR）超固结比
overconsolidated soil 超固结土
particle size distribution of soils 颗粒级配
passive earth pressure 被动土压力
peak strength 峰值强度
permeability test 渗透试验
permeability 渗透性
phreatic line 浸润线
pile caps 承台
pile cushion 桩垫
pile driving（by vibration）（振动）打桩
pile foundation 桩基础
pile groups 群桩
pile head＝butt 桩头
pile plan 桩位布置图
pile pulling test 拔桩试验
pile rig 打桩机
pile spacing 桩距
piping 管涌
plastic limit test 塑限试验
plasticity index 塑性指数
plate loading test 静力荷载试验
poorly-graded soil 级配不良土
pore pressure dissipation test 孔隙水压力消散试验
pore water pressure 孔隙水压力
pore-pressure distribution 孔压分布
porosity 孔隙率
precast concrete pile 预制混凝土桩
precast reinforced concrete piles 钢筋混凝土预制桩
preconsolidation pressure 前期固结压力
preloading 预压法
presoaking 预浸水法
pressuremeter test 旁压试验

prestressed concrete pipe pile 预应力混凝土管桩
primary consolidation 主固结
principal plane 主平面
principle of effective stress 有效应力原理
quick direct shear test 快剪试验
rammed-cement-soil pile 夯实水泥土桩法
Rankine's earth pressure theory 朗肯土压力理论
rate of consolidation 固结速率
rebound curve 回弹曲线
reciprocating activity 往返活动性
recompression curve 再压缩曲线
reduced load 折算荷载
reinforced concrete sheet pile 钢筋混凝土板桩
reinforced earth bulkhead 加筋土挡墙
reinforced earth 加筋土
reloading 再加载
replacement ratio（复合地基）置换率
residual strength 残余强度
resonant column test 共振柱试验
retaining wall 挡土墙
rigid foundation 刚性基础
rigid plastic model 刚塑性模型
ring shear test 环剪试验
routine consolidation test 常规固结试验
safety factor of slope 边坡稳定安全系数
sand boiling 砂沸
sand drain 砂井
sand relative density test 砂的相对密实度试验
sandy silt 砂质粉土
saturated density 饱和密度
saturated soil 饱和土
saturated unit weight 饱和重度
secant modulus 割线模量
secondary consolidation 次固结
seepage discharge 渗流量
seepage failure 渗透破坏
seepage force 渗透力

seepage pressure 渗透压力
seepage velocity 渗流速度
seepage 渗透（流）
seismic predominant period 地震卓越周期
self weight collapse loess 自重湿陷性黄土
sensitivity of cohesive soil 黏性土的灵敏度
settlement 沉降
shaft 竖井；桩身
shallow foundation 浅基础
shallow treatment 浅层处理
shear modulus 剪切模量
shear strain rate 剪切应变速率
shear strength 抗剪强度
shear stresses 剪应力
sheet pile structure 板桩结构
sheet pile 板桩
sheet pile-braced cuts 板桩围护
shell foundation 壳体基础
short-term transient load 短期瞬时荷载
shrinkage limit test 缩限试验
silent piling 静力压桩
silt 粉土
silty clay 粉质黏土
single-grained structure 单粒结构
slice method 条分法
slip line 滑动线
slope stability analysis 土坡稳定分析
slope 土坡
soft clay ground 软土地基
soft clay 软黏土
soft soil 软土
soil dynamics 土动力学
soil mass 土体
soil mechanics 土力学
specific gravity test 比重试验
specific gravity 比重
spread footing 扩展基础
square spread footing 方形独立基础
stability of foundation soil 地基稳定性

stability of retaining wall 挡土墙稳定性
standard penetration test（SPT）标准贯入试验
state of elastic equilibrium 弹性平衡状态
state of limit equilibrium 极限平衡状态
steel pile 钢桩
steel pipe pile 钢管桩
steel sheet pile 钢板桩
strain control triaxial compression apparatus 应变控制式三轴压缩仪
strain harding law 加工硬化定律
stress control triaxial compression apparatus 应力控制式三轴压缩仪
stress path 应力路径
stress ratio of liquefaction 液化应力比
stress wave in soils 土中应力波
strip foundation 条形基础
surcharge preloading 超载预压法
surface compaction 表层压实法
surface wave test（SWT）表面波试验
Swedish circle method 瑞典圆弧滑动法
swelling index 回弹指数
tangent modulus 切线模量
Terzaghi bearing capacity theory 太沙基承载力理论
Terzaghi's consolidation theory 太沙基固结理论
time factor 时间因子
tip resistance 桩端阻力
total stress approach of shear strength 抗剪强度总应力法
total stress failure envelope 总应力破坏包线
total stress strength parameter 总应力强度参数
total stress 总应力
triaxial compression test 三轴压缩试验
tri-phase soil 三相土
true triaxial apparatus 真三轴仪
twin shear stress yield criterion 双剪应力屈服模型

ultimate bearing capacity of foundation soil 地基极限承载力
unconfined compression strength test 无侧限抗压强度试验
unconsolidated-undrained triaxial test 不固结不排水试验
underconsolidated soil 欠固结土
underlying soil 下卧层
undrained shear strength 不排水抗剪强度
uniaxial tension test 单轴抗拉强度
uniformity coefficient 不均匀系数
uniformity coefficient 不均匀系数
unloading 卸载
unsaturated soil 非饱和土
uplift pile 抗拔桩
vacuum preloading 真空预压法
vane shear strength 十字板抗剪强度
vane shear test 十字板剪切试验
vertical allowable load capacity of single pile 单桩竖向抗压容许承载力
vertical ultimate carrying capacity of single pile 单桩竖向抗压极限承载力
vertical ultimate load capacity of pile groups 群桩竖向极限承载力
vertical ultimate uplift resistance of single pile 单桩竖向抗拔极限承载力
vibration isolation 隔振
vibro replacement stone column 振冲碎石桩
vibro-densification，compacting 振密、挤密法
virgin compression curve 原始压缩曲线
viscoelastic model 黏弹性模型
viscous damping 黏滞阻尼
void ratio 孔隙比
volumetric deformation modulus 体积变形模量
watercontent test 含水量试验
wave velocity method 波速法
well point system 井点系统（轻型）
well-graded soil 级配良好土
work hardening 加工硬化
yield surface 屈服面

附录 D 结构动力学

acceleration impedance 加速度阻抗
acceleration mobility 加速度导纳
accelerograms 加速度图；加速度时程
amplitude spectrum 幅值谱
amplitude 振幅
amplitude-frequency characteristics 幅频特性
analysis of frequency 频率分析
angular frequency 角频率
approximate frequency analysis 近似频率分析
autocorrelation function 自相关函数
auto-covariance function 自协方差函数
auto-spectral density 自功率谱密度
average absolute value 平均绝对值
axial force effects 轴力效应
axial vibration modes 轴向振动振型
axial wave propagation 轴向波传播
boundary conditions 边界条件
buckling analysis 屈曲分析
circular undamped frequency 无阻尼圆频率
coherence function 相干函数
combined stiffness 联合刚度
complex amplitude 复数振幅
complex frequency response 复频反应
complex mode 复模态
complex-stiffness 复刚度
consistent mass matrix 一致质量矩阵
constant average acceleration method 常平均加速度法

convolution integral 卷积积分
correlation coefficient 相关系数
coupled system 耦合体系
critical buckling load 临界屈曲荷载
critical damping 临界阻尼
cross-spectral density 互功率谱密度
cross-spectral density 互谱密度
D'Alembert's principle 达朗伯原理
damped natural frequency 阻尼固有频率
damped response 有阻尼反应
damping exponent 衰减指数
damping ratio 阻尼比
damping 阻尼
degrees of freedom 自由度
discrete coordinate systems 离散坐标体系
discrete loading 随机荷载
discrete spectrum 离散谱
displacement impedance 位移阻抗
distributed parameter system 分布参数体系
distribution function 分布函数
Duhamel integral 杜哈梅积分
dynamic loading 动力荷载
dynamic magnification factor 动力放大系数
dynamic matrix 动力矩阵
dynamic stiffness 动力刚度
earthquake motion 地震动
earthquake structural response 地震结构反应
effective modal mass 等效模态质量
eigenproblem 特征问题
energy spectrum density 能量谱密度
equation of motion 运动方程
equations of constraint 约束方程
excitation 激励
fast fourier transform 快速傅里叶变换
flexibility 柔度
forced response 强迫响应
forced vibration 强迫振动
foundation-structure interaction 基础-结构相互作用

free field motion 自由场运动
free response 自由响应
free vibrations 自由振动
free-vibration decay 自由振动衰减
frequency domain analysis 频域分析
frequency equation 频率方程
frequency response function 频响函数
generalmass matrix 一般质量矩阵
generalized coordinates 广义坐标
generalized damping 广义阻尼
generalized displacement 广义位移
generalized load 广义荷载
generalized mass 广义质量
generalized properties 广义特性
generalized SDOF system 广义单自由度体系
generalized stiffness 广义刚度
geometric stiffness 几何刚度
ground accelerations 地面加速度
harmonic loading 简谐荷载
harmonic vibration 简谐振动
impedance functions 阻抗函数
impedance matrix 阻抗矩阵
impulse loading 脉冲荷载
impulse response function 冲激响应函数
impulsive loads 冲击荷载
incident waves 入射波
inertial force 惯性力
initial phase 初相位
initial shock response spectrum 冲击初始响应谱
interpolation functions 插值函数
inverse iteration 逆迭代
inversematrix iteration 逆矩阵迭代
Jacobian transformation 雅克比变换
linear acceleration method 线性加速度法
lumpedmass matrix 集中质量矩阵
lumped mass 集中质量
magnification factor 放大系数
marginalprobability 边缘概率

mass influence coefficients 质量影响系数
mass matrix 质量矩阵
mass moment of inertia 质量惯性矩
matrix iteration with shifts 移位逆矩阵迭代
matrix iteration 矩阵迭代
matrix of FRF 频响函数矩阵
matrix of modal shape 振型矩阵
matrix of transfer function 传递函数矩阵
maximum shock response spectrum 冲击最大响应谱
MDOF systems 多自由度体系
modal analysis 模态分析
modal coordinates 模态坐标
modal damping coefficient 模态阻力系数
modal damping ratio 模态阻尼比
modal frequency 模态频率
modal impedance 模态阻抗
modal loads 模态荷载
modal mass 模态质量
modal matrix 模态矩阵
modal parameter 模态参数
modal participation factors 振型参与系数
modal shape 模态振型
modal stiffness 模态刚度
mode shape 振型
mode superposition 振型叠加
mode 模态
multiple excitation 多点激励
natural frequency 固有频率
newmark beta methods 纽马克 β 法
nonlinear response analysis 非线性反应分析
nonstationary random process 非平稳随机过程
nonstationary 非平稳的
normal coordinate 正则坐标
normal mode equations 正则振型方程
normal mode 实模态
normalizing mode shapes 正则化振型
orthogonality 正交阻尼，正交性

overdamped 过阻尼
peak-value 峰值
period 周期
periodic loading 周期荷载
periodic vibration 周期振动
phase angle 相位角
phase frequency characteristics 相频特性
phase spectrum 相位谱
power spectral density 功率谱密度
principal frequency 主频率
proportional viscous damping 比例黏性阻尼
pseudo-acceleration response spectrum 伪加速度反应谱
pseudo-velocity response spectrum 伪速度反应谱
P-wave，S-wave P 波，S 波
quasi-static displacements 拟静力位移
random loading 随机荷载
random processes 随机过程
Rayleigh-Ritz method 瑞利-里兹法
rectangular impulse 矩形脉冲
refracted wave 折射波
resonance 共振
resonant amplification 共振放大
response integral 反应积分
response ratio 反应比
response spectrum 反应谱
rotational excitation 转动激励
selection of shape 振型选择
shock excitation 冲击激励
shock response spectrum 冲击响应谱
shock response 冲击响应
shock spectra 震动谱
spectral acceleration 谱加速度
spectral displacement 谱位移
step-by-step integration 逐步积分
stochastic response 随机反应
strong ground motion 强地面运动
strong motion accelerogram

强烈地震加速度图
strong motion earthquake 强震
structural damping 结构阻尼
structural-property matrices 结构特性矩阵
support excitation 支座激励
time domain 时域
transfer function relationships 传递函数关系
transient process 瞬态过程
transient response 瞬态反应
translational excitation 平动激励
triangular impulse 三角形脉冲
uncoupling 解耦

undamped modal frequency 无阻尼模态频率
underdamped 低阻尼
vector of mode shape 振型矢量
velocity impedance 速度阻抗
vibration isolation 隔振
vibration process 振动过程
vibration system 振动系统
virtual displacements 虚位移
virtual work 虚功
wave velocity 波速
white noise 白噪声
Wilson θ method 威尔逊θ法

附录 E 理 论 力 学

acceleration 加速度
aerodynamics 空气动力学
amplitude 振幅
angular momentum 动量矩
angular-impulse 角冲量
angular-momentum 角动量
angular-velocity 角速度
axis 轴
bearing 轴承，支撑面
bolt 螺栓
cam 凸轮
cantilever 悬臂
cast-iron 铸铁
center-of-gravity 重心
center-of-mass 质心
centripetal force 向心力
centroid 形心
circular-frequency 圆频率
clockwise (CW) 顺时针
collar 套筒
combine motion 复合运动
component 分量，构成元件
composite-body 组合体

composite-motion 复合运动
concurrent 汇交的
conservation-of-momentum 动量守恒
conservative-force 保守力
constraint 约束
contour 等高线，参照线
conventional 惯例的
convert-conversion 转化
Coriolis-acceleration 科氏加速度
Coulomb's-law-of-friction 库仑摩擦定律
counterclockwise (CCW) 逆时针
couple (s) 力偶
crank 曲柄
cross-product 叉乘法
cycloid 摆线
cylinder 圆柱，汽缸
damped-vibration 衰减振动
damp 阻尼，衰减的
deformation 形变
degrees-of-freedom 自由度
density 密度
direction-cosine 方向余弦
displacement 位移

distributed-load 分布载荷
dot-product 点乘法
dynamics 动力学
eccentricity 偏心距，离心率
elongation（弹簧等）伸长量
equation-of-motion 运动方程
equilibrium 平衡
equipotential-surfaces 等势面
finite element method 有限元方法
frequency 频率
friction 摩擦
graphically 图解法
gravitation 引力
gravity 重力
hinge 门纹，铰链
horizontal 水平的
humidity 湿度
impact 碰撞
impending 临界的
impulse 冲量
incline 倾斜
indicate＝locate 标明
inertia 惯性，惯量
inertial-reference-frame 惯性系
initial-condition 初始条件
initial-initially 初始的
instant 瞬时
jack 千斤顶
joint＝node 结合，节
joule 焦耳
key 键，键槽
kinematics 运动学
kinetic-energy 动能
kinetics＝dynamics 动力学
linear-vibration 线振动
load 载荷
load-intensity 载荷强度
magnitude 量值大小
mass 质量

mechanical-energy 机械能
mechanics 力学
mild-steel 低碳钢
misalignment 未对准
moment 矩
moment-arm 矩臂
momentum 动量
natural-frequency 固有频率
non-inertial-reference-frame 非惯性系
normal 法向的
numerical 数值的
nut 螺母
operator 计算符，算子
orthogonal-component 正交分量
parallel-axis-theorem 平行轴定理
parallelogram-law 平四法则
particle 质点
pedal 踏板
pendulum 摆
pitch 螺距
plane 平面
plank 铺板
position-vector 位矢
potential-energy 势能
principle 原理
principle-of-change-of-momentum 动量定理
principle-of-work-and-energy 动能定理
procedure＝step 步骤
projection 投影
property 性质
pulley 滑轮
radius-of-gyration 回转半径
rate-of-change 变化率
rectangular-component 正交分量
rectilinear 直线运动
repel 排斥
resistance 阻力
resonance 共振
restoring-force 回复力

restrict-restriction 约束
resultant moment 合力矩
resultant 合力
rigid body 刚体
rotate-rotation 旋转
scale 天平，磅秤
screw 螺丝
screwdriver 螺丝刀
section 部件，截面
sector 扇形
self lock 自锁
shaft 连杆，轴
simple-pendulum 单摆
skid＝brake 制动
slack 松弛，缝隙
slope 斜度，斜率
slot 滑槽
socket 插槽，嵌槽
speed 速率（s）
spool 线框，线轴
stability 稳定性
statics 静力学
stiffness 刚度

survey 测量，调查
suspend 悬挂
term 术语
theorem 定理法则
thread 螺纹
tire＝tyre 轮胎
torque 扭矩
traction 牵引
trajectory 轨迹
translate 平动
tripod 三脚架
truss 桁架
universal-joint 万向节
validate 验证（有效）
velocity 速度（v）
versus 对，比
virtual-work 虚功
vise 虎钳
watt 瓦特
wear 磨损
wedge 楔
weld 焊接
winch 绞盘

附录F 材料力学

active strain gage 工作应变计
allowable load method 许用荷载法
allowable load 许用荷载
allowable stress method 许用应力法
allowable stress 许用应力
angel of rotation 转角
angel of twist 扭转角
approximately differential equation of the deflection curve 挠曲轴近似微分方程
average stress 平均应力
axial compression 轴向压缩
axial deformation 轴向变形

axial force diagram 轴力图
axial force 轴力
axial rigidity 拉压刚度
axial tension 轴向拉伸
axially loaded bar 拉压杆，轴向承载杆
bar 杆，杆件
beam of constant strength 等强度梁
beam of variable cross section 变截面梁
beam 梁
bearing stress 挤压应力
bending moment diagram 弯矩图
bending moment 弯矩

bending 弯曲
body force 体积力
boundary conditions 边界条件
bridge balancing 电桥平衡
brittle materials 脆性材料
buckling 失稳
centroidal axis 形心轴
combined deformation 组合变形
compatibility equation 变形协调方程
compensating block 补偿块
composite area 组合截面
composite material 复合材料
constraint torsion 约束扭转
continuity condition 连续条件
continuous beam 连续梁
core of section 截面核心
critical load 临界荷载
critical stress 临界应力
cyclic stress 交变应力，循环应力
deflection curve 挠曲轴
deflection 挠度
deformation 变形
degree of a statically indeterminate problem 静不定次，超静定次数
density of energy of volume change 体积改变能密度
displacement method 位移法
distortion energy theory 畸变能理论
distortional strain energy density 畸变能密度
distributed force 分布力
ductile materials 塑性材料，延性材料
dynamic load 动荷载
eccentric compression 偏心压缩
eccentric tension 偏心拉伸
elastic deformation 弹性变形
elastic-perfectly plastic assumption 理想弹塑性假设
elongation 伸长率
endurance limit 疲劳极限，条件疲劳极限

equation of bending moment 弯矩方程
equation of deflection curve 挠曲轴方程
equation of shear force 剪力方程
equivalent length 有效长度
equivalent stress 相当应力
Euler's formula 欧拉公式
experimental stress analysis 实验应力分析
fatigue life 疲劳寿命
fatigue rupture 疲劳破坏
flexural rigidity 弯曲刚度
force method 力法
frame 刚架，构架
free torsion 自由扭转
full bridge 全桥接线法
gage length 标距
generalized Hook's law 广义胡克定律
geometrical properties of an area 截面几何性质
half bridge 半桥接法
high-cycle fatigue 高周疲劳
Hook's law for shear 剪切胡克定律
Hook's law 胡克定律
impact load 冲击荷载
initial stress 初应力，预应力
intermediate columns 中柔度杆
internal forces 内力
isotropical material 各向同性材料
lateral deformation 横向变形
limit load 极限荷载
linear elastic body 线弹性体
long columns 大柔度杆
low-cycle fatigue 低周疲劳
maximum shear stress theory 最大切应力理论
maximum stress 最大应力
maximum tensile strain theory 最大拉应变理论
maximum tensile stress theory 最大拉应力理论
mechanical properties 力学性能

mechanics of materials 材料力学
method of sections 截面法
middle plane 中面
Miner's law 迈因纳定律
minimum stress 最小应力
modulus of elasticity 弹性模量
Mohr theory of failure 摩尔强度理论
Mohr's circle for stresses 应力圆，摩尔圆
moment of inertia 惯性矩
necking 颈缩
neutral axis 中性轴
neutral surface 中性层
normal strain 正应变
normal stress in bending 弯曲正应力
normal stress 正应力
offset yielding stress 名义屈服强度
parallel axis theorem 平行移轴定理
percentage reduction in area 断面收缩率
plane bending 平面弯曲
plane cross-section assumption 平面假设
plastic deformation 塑性变形，残余变形
plastic hinge 塑性铰
Poisson's ratio 泊松比
polar moment of inertia 极惯性矩
primary system 基本系统
principal axis 主轴
principal centroidal axis 主形心轴
principal centroidal moments of inertia
   主形心惯性矩
principal moment of inertia 主惯性矩
principal planes 主平面
principal stress trajectory 主应力迹线
principal stress 主应力
product of inertia 惯性积
proportional limit 比例极限
pure bending 纯弯曲
pure shear 纯剪切
radius of gyration of an area 惯性半径
reciprocal-displacement theorem 位移互等定理

reciprocal-work theorem 功的互等定理
redundant restraint 多余约束
resistance strain gage 电阻应变计
resistance strain indicator 电阻应变仪
rigid joint 刚结点
safety factor for stability 稳定安全因数
safety factor 安全因数
Saint-Venant's principle 圣维南原理
section modulus in bending 抗弯截面系数
section modulus in torsion 抗扭截面系数
sensitive grid 敏感栅
shape factor 形状系数
shear center 弯曲中心
shear flow 剪流
shear force diagram 剪力图
shear force 剪力
shear modulus 切变模量
shear strain 切应变
shear stress in bending 弯曲切应力
shear stress 切应力
short columns 小柔度杆
slenderness ratio 细长比，柔度
slip-line 滑移线
spring constant 弹簧常量
stability condition 稳定条件
stability 稳定性
state of biaxial stress 二向应力状态
state of plane stress 平面应力状态
state of stress 应力状态
state of triaxial stress 复杂应力状态
state of triaxial stress 三向应力状态
static load 静荷载
static moment 静矩，一次矩
statically determinate problem 静定问题
statically indeterminate problem
   静不定问题，超静定问题
stiffness 刚度
strain energy density 应变能密度
strain energy 应变能

strain gage 应变计
strain rosette 应变花
strength condition 强度条件
strength 强度
stress amplitude 应力幅
stress concentration factor 应力集中因数
stress concentration 应力集中
stress ratio 应力比
stress ratio 应力速率
stress 应力
stress-cycle curve 应力寿命曲线
stress-strain diagram 应力应变图
structural member 构件
superposition method 叠加法
superposition principle 叠加原理
theorem of conjugate shearing stress 切应力互
　　　　　　　　　　　　　　　　等定理
theory of strength 强度理论
thermal stress 热应力
thin-walled tubes 闭口薄壁杆
three-element delta rosette 三轴等角应变花

three-element rectangular rosette 三轴直角
　　　　　　　　　　　　　　　　应变花
torque diagram 扭矩图
torsion 扭转
torsional moment 扭矩
torsional rigidity 扭转刚度
transformation equation 转轴公式
twisting moment 扭力矩
ultimate stress in tension 抗拉强度
ultimate stress in torsion 抗扭强度
ultimate stress in torsion 扭转极限应力
ultimate stress 极限应力
uniaxial stress 单向应力，单向受力
unit couple 单位力偶
unit load 单位荷载
unit-load method 单位荷载法
volume strain 体应变
yield s trength 屈服强度
yield 屈服
yielding stress in torsion 扭转屈服强度

附录 G　结 构 力 学

accessory part 附属部分
amplitude of vibration 振幅
antisymmetrical load 反对称荷载
arch 拱
assembled stiffness equation 整体刚度方程
assembled stiffness matrix 整体刚度矩阵
axial force 轴力
beam 梁
bending moment 弯矩
carry-over factor 传递系数
characteristic equation 特征方程
circular frequency 圆频率
combined truss 联合桁架
complicated truss 复杂桁架

composite structure 组合结构
connection link 链杆
constraint or restraint 约束
continuous beam 连续梁
critical damping coefficient 临界阻尼常数
critical state 临界状态
damping 阻尼
damping ratio 阻尼比
degree of freedom 自由度
degree of indeterminacy 超静定次数
displacement 位移
displacement coefficient 位移系数
displacement method 位移法
distribution factor 分配系数

dynamic coefficient 动力系数
dynamic load 动力荷载
element analysis 单元分析
element end displacement matrix 单元杆端位矩阵
elementend force matrix 单元杆端力矩阵
element localization vector 单元定位向量
element stiffness equation 单元刚度方程
element stiffness matrix 单元刚度矩阵
element 单元
energy method 能量法
envelope of bending moment 弯矩包络图
envelope of internal force 内力包络图
equation of force method 力法方程
equivalent nodal load 等效结点荷载
first frequency 第一频率
first mode shape 第一振型
fixed support 固定支座
fixed-end moment 固定弯矩
fixed-end shear force 固定剪力
flexibility coefficient 柔度系数
flexibility matrix 柔度矩阵
flexibility method 柔度法
force method 力法
forced vibration 强迫振动
frame 刚架
free vibration 自由振动
frequency equation 频率方程
frequency 频率
frequentation unstable system 常变体系
fundamental frequency 基本频率
fundamental mode shape 基本振型
fundamental part 基本部分
fundamental structure 基本结构
fundamental system 基本体系
fundamental unknown 基本未知量（力）
generalized displacement 广义位移
generalized force 广义力
geometrically stable system 几何不变体系

geometrically unstable system 几何可变体系
global analysis 整体分析
harmonic load 简谐荷载
hinged joint 铰结点
hinged support 铰支座
impulsive load 冲击荷载
infinite degree of freedom system 无限自由度体系
influence factor 影响系数
influence line 影响线
influence line of internal force 内力影响线
instantaneous unstable system 瞬变体系
internal force 内力
joint load 结点荷载
joint or node 结点
joint rotation displacement 结点角位移
kinematics method 机动法
line stiffness 线刚度
load 荷载
mass matrix 质量矩阵
matrix analysis of structures 结构矩阵分析
matrix displacement method 矩阵位移法
method of graph multiplication 图乘法
method of joint 结点法
method of lumped mass 集中质量法
method of moment distribution 力矩分配法
moving load 移动荷载
multi-span statically determinate beam 多跨静定梁
necessary constraint 必要约束
normal mode shape 主振型（振型）
normalized mode shape 标准化主振型
orthogonality of normal modes 主振型的正交性
period load 周期荷载
period 周期
plane truss 平面桁架
plate and shell structures 板壳结构
post treatment method 后处理法

pretreatment method 先处理法
principle of virtual force 虚力原理
principle of virtual work 虚功原理
push force 推力
random load 随机荷载
Rayleigh method 瑞利法
reaction force coefficient 反力系数
reasonable axis of arch 合理拱轴
reciprocal theorem of displacement 位移互等定理
reciprocal theorem of work 功的互等定理
resonance 共振
rigid arm 刚臂
rigid body 刚体
rigid frame with sideway 有侧移刚架
rigid frame without sideway 无侧移刚架
rigid joint 刚结点
roller support 滚轴支座
rotational stiffness 转动刚度
self-internal force 自内力
shape function 形状函数
shear force 剪力
simple hinge 单铰
simple rigid joint 单刚结点
simple truss 简单桁架
single degree of freedom system 单自由度体系
slope-deflection equation 转角位移方程
spectrum of magnification factor 动力系数反应谱
static load 静力荷载
static method 静力法
statically determinate structure 静定结构
statically undeterminate structure 超静定结构
staticallydeterminate plane frame 静定平面刚架
stiffness coefficient 刚度系数
stiffness equation 刚度方程
stiffness matrix 刚度矩阵
stiffness method 刚度法
structural compute diagram 结构的计算简图
structure 结构
structure of bar system 杆件结构
suddenly applied constant load 突加荷载
superfluous constraint 多余约束
symmetrical load 对称荷载
symmetrical structure 对称结构
symmetry 对称性
the most unfavorable load position 最不利荷载位置
three-hinged arch 三铰拱
transformation of coordinates 坐标转换
truss 桁架
unit load method 单位荷载法
virtual hinge 虚铰
viscous damping force 黏滞阻尼力
zero bar 零杆

附录 H 基 本 术 语

mathematics 数学
number 数字
even number 双数
odd number 单数
calculate 计算
calculate mentally 口算
calculate using pen-and-paper 笔算
vertical form 竖式
1-digit number 一位数
2-digit number 两位数
word problem 文字题
storyproblem 应用题

附录 I 常用符号

+ plus
− minus
= equal(s)

\> is greater/more than
< is less than
( ) brackets

附录 J 数学知识

absolute value 绝对值
addend 加数
addition 加法
algebra 代数
analysis 解析
approach 途径，趋近，方法
approximate number 近似数
arc-coordinates 弧坐标
augend 被加数
average 平均数
cardinal number 基数
Cartesian-coordinatesystem 笛卡儿坐标系
coefficient 系数
coefficient 系数
complex number 复数
constants 常数
cross-product 叉乘法
cube root 立方根
cube 三次方
cubic equation 三次方程
decimal point 小数点
decimal 小数
definition 定义
denominator 分母
derivative 导数
derivative 导数
determinant 行列式
difference 差
differential 微分

differential 微分
differential-equation 微分方程
dimension 维，量纲，度量单位
direction-cosine 方向余弦
discount 折扣
divide 除
dividend 被除数
division 除法
divisor 除数
dot-product 点乘法
equal-sign 等号
equation 等式；方程式
equivalence 等价
equivalent 等同的
even number 偶数
exponent 指数
extracting roots 开方
formula 公式
fourier-series 傅立叶级数
fraction 分数
function 函数
general solution 通解
gradient 梯度
homogeneous 齐次的
hundreds 百
identity 恒等式
imaginary number 虚数
inequality 不等式
infinitesimal 无穷小

initial-condition 初始条件	particular-solution 特解
integer number 整数	percent 百分比
integral 积分	period 周期
integral 积分	phase 相位
interest 利息	place value chart 数位表
interval 区间	plot 图像
inverse 倒数	plus sign 加号
invert 反解	plus 加上，正的
irrational number 无理数	plus；add；and；increase 加
is；equal 等	polar-coordinate 极坐标
lemma 引理	polynomial 多项式
logarithms 对数	positive integer number 正整数
matrix 矩阵	positive 正
matrix 矩阵	power 幂；乘方
minuend 被减数	product 积
minus sign 减号	proportion 比例
minus 减去，负的	proportional 成比例的
minus；decrease；subtract 减	quadratic equation 二次方程
monomial 单项式	quadruplicate 乘四次方
multiplicand 被乘数	quotient 商
multiplication 乘法	radical sign 根号
multiplication 乘法	rate-of-change 变化率
multiplier 乘数	ratio 比
multiply；multiplied by；times 乘	rational number 有理数
natural number 自然数	real number 实数
natural system of coordinates 自然坐标系	recurring decimal 循环小数
necessary and sufficient condition 充要条件	remainder 余数
negative integer number 负整数	round off 舍入
negative 负	scalar 标量
non-homogeneous 非齐次的	second-order-differentiation 二阶微分
nought point four 零点四	sign of equality 等号
numerator 分子	sign of inequality 不等号
numerical 数值的	significant number 有效数
odd number 奇数	simple equation 一次方程
operator 计算符，算子	slope 斜度，斜率
ordinal number 序数	square root 平方根
ordinary differential 常微分	subscript 下标
parameter 参数	subtraction 减法
partial-differentiation 偏微分	subtraction 减法

subtrahend 减数
sum 和
summation 和
superposition 叠加
taylor-series 泰勒级数
tens 十
theorem 定理

three cubed 三次方的
to rise to the power of five 使乘五次方
units/ones 个
unknown number 未知数
variable 变量
X squared 某数的平方
zero; naught; 0 零

附录K 图形名称

line 线
angle 角
intersecting line 相交线
parallel line 平行线
triangle 三角形
quadrilateral 四边形
rectangle 矩形
lozenge 菱形
polygon 多边形
arc 弧
perimeter 周长
area 面积
diameter 直径
volume 体积
cuboid 长方体
cube 立方体
sphere 球
rectangle 长方形
square 正方形
circle 圆
side 边
angle 角
face 面
cone 圆锥
cylinder 圆柱
sector 扇形

conic-section 圆锥曲线
ellipse 椭圆
hyperbola 双曲线
parabolic 抛物线
cycloid 摆线
eccentricity 偏心距，离心率
helix-helical 螺旋
line-segment 线段
projection 投影
radii＝radius 半径
right-angle 直角
vertex-angle 顶角
plane 平面
section 截面
diagonal 对角线
centroid 形心
symmetry 对称
curvature 曲率
curved-surface 曲面
law-of-cosine 余弦定理
law-of-sine 正弦定理
perpendicular 垂直的
tangent-tangential 切向（的）
coplanar 共面的
non-collinear 不共线的
non-coplanar 不共面的

附录 L 常用数学表达式

1/2: a half; one half

1/3: a third; one third

2/3: two thirds

1/4: a quarter; one quarter; a fourth; one fourth

1/100: a (one) hundredth

1/1000: a (one) thousandth

113/324: one hundred and thirteen over three hundred and twenty-four

4 2/3: four and two-thirds

45 89/23: forty-five and eighty-nine over twenty-three

0.1: one tenth; point one

0.01: one hundredth; point zero one

0.001: one thousandth; point zero zero one; point two zero one

2050.0357: two thousand and fifty point zero three five seven

0.25: zero point two five; point two five repetend five; zero point two five recurring

对 483579 四舍五入到千位: round off 483579 to nearest thousand

10^8: one followed by eighteen zeros

-30.8: negative thirty point eight

2-3i: two minus three i; two minus three times i

∞: infinity

$x+y=z$: x plus y is z; add x to y is z; x and y is z

$(x+y)$: bracket x plus y bracket closed

$x-y$: x minus y; subtract y from x; y from x; x subtracts y

$x \pm y$: x plus or minus y

$x \times y$; $x \cdot y$; xy: multiply x by y; x multiplied by y; x by y; x times y;

$x \div y$: divide x by y; y into x; x over y; $x:y$; the ration of x to y

$x \propto y$: x varies as y; x is in direct proportion to y

$x=y$: x equals y; x is equal to y; x is y

$x \neq y$: x is not equal to y; x is not y

$x \equiv y$: x is identical to y; x is equivalent to y; x is equivalent to y

$x \approx y$: x is approximately equal to y; x approximately equals y

$x > y$: x is greater than y; x is more than y

$x \gg y$: x is much greater than y; x is far greater than y

$x \geq y$: x is greater than or equal to y

$x < y$: x is less than y

$x \ll y$: x is much less than y

$x \leq y$: x is less than or equal to y

$0 < x < 1$ zero is less than x is less than 1; x is greater than zero and less than 1

$0 \leq x \leq 1$ zero is less than or equal to x is less than or equal to 1

x^2: x square; x squared; the square of x; the second power of x; x to second power

x^3: x cube; x cubed; the cube of x; the third power of x; x to the third power

x^n the nth power of x; x to the nth power; x to the power n

$\sqrt{x}$ the square root of x

$\sqrt[3]{x}$ the cube root of x

$\sqrt[n]{x}$ x to the power one over n; the nth root of x

%: per cent

2%: two per cent
$\%_0$: per mill
$5\%_0$: five per mill
$\log_n x$: log x to the base n
$\log_{10} x$: log x to the base 10, common logarithm
$\ln x$: log x to the base e; Natural (Napierian) logarithm
$\exp(x)$: exponential function of x; e to the power x
x^n: x to the power n; the nth power of x
x^{-n}: x to the (power) minus n
$(x+y)^2$: x plus y all squared
$(x/y)^2$: x over y all squared
x_i: x subscript i; x suffix i; x sub i
e^x; $\exp(x)$: exponential function of x, e to the power x
Σ: the summation of
Π: the product of
$\sum_{i=1}^{n} x_i$: the summation of x sub i, where i goes from 1 to n; the sum from i equals one to n x_i; the sum as i runs from one to n of the x_i
$\prod_{i=1}^{n} x_i$: the product of x sub i, where i goes from 1 to n; the product of all x_i from i equals one to n; the product of all x_i from i equals one to infinity
$|x|$: the absolute value of x; modulus x
$\bar{x}$: the mean value of x; x bar; x hat; x tilde
x^*: x asterisk; x prime; x double prime
$\{\ \}$: braces
$(\ )$: round brackets; parenthesis
$[\ ]$: square (angular) brackets
$x \to \infty$: x approaches infinity
$\cong$: be congruent to
$\sim$ or $\backsim$: be similar to

$5 \times (6-3)$: the product of five and six decreased by three
$5 \times 6 - 3$: the product of five and six, decreased by three
b': b prime
b'': b second prime, b double prime
b''': b triple prime
$g(x)$: function g of x
Δx: the increment of x
Δ: finite increment or difference
dx: dee x; dee of x; differential x
$\dfrac{dy}{dx}$: the differential coefficient of y with respect to x; the first derivative of y with respect to x
$\dfrac{d^2 y}{dx^2}$: the second derivative of y with respect to x
$\dfrac{d^n y}{dx^n}$: the nth derivative of y with respect to x
$\dfrac{\partial y}{\partial x}$: the partial derivative of y with respect to x, where y is a function of x and another variable (or other variables)
∇: del; nabla
∇^n: nth del (nabla)
$\int$: integral of
$\iint$: double integral of
$\int \cdots \int$: N-fold integral of
$\int_a^b$: integral between limits a and b
$\vec{F}$: vector F
a_2: a sub 2
$20°$: twenty degree
$7'$: seven minute, seven feet
$13''$: thirteen seconds, thirteen inches
$0°C$: zero degree centigrade

$32°F$: thirty-two degrees fahrenheit

$f(x)$: fx; f of x; the function f of x

$\lim_{x \to 0}$: the limit as x approaches 0

$\lim_{x \to +0}$: the limit as x approaches 0 from above

$\lim_{x \to -0}$: the limit as x approaches 0 from below

$\exists$: there exists

$\forall$: for all

$\{\ \}$; ϕ: empty set

$f'(x)$: f prime x; f dash x; the (1st) derivative of f with respect to x

$f''(x)$: f double-prime x; f double-dash x; the second derivative of f with respect to x

$f'''(x)$: f triple-prime x; f triple-dash x; the third derivative of f with respect to x

$f^4(x)$: four x; the fourth derivative of f with respect to x

$x!$: x factorial

Δ: finite difference or increment

Δx or δx: the increment of x

$x \Rightarrow y$: x implies y; if x, then y

$x \Leftrightarrow y$: x if and only if y; x is equivalent to y; x and y are equivalent

$x \in A$: x belongs to A; x is an element (or a member) of A

$x \notin A$: x does not belong to A; x is not an element (or a member) of A

$A \subset B$: A is contained in B; A is a subset of B

$A \supset B$: A contains B; B is a subset of A

$A \cap B$: A cap B; A meet B; A intersection B

$A \cup B$: A cup B; A join B; A union B

$A - B$: A minus B; the difference between A and B

$A \times B$: A cross B; the cartesian product of A and B (A 与 B 的笛卡尔积)

$\|A\|$: the norm (or modulus) of A

AB: the length of the segment AB

A^T: A transpose; the transpose of A

A^{-1}: A inverse; the inverse of A

$x \to y$: x maps into y; x is sent (or mapped) to y

sin: sine

cos: cosine

tan: tangent

cot: cotangent

sec: secant

csc: cosecant

arcsin: arcsine

arcos: arccosine

sinh or sh: the hyperbolic sine

cosh or ch: the hyperbolic cosine

附录 M 常用希腊字母

大写	小写	英文注音	国际音标注音	中文读音	意义
A	α	alpha	[aːlf]	阿尔法	角度；系数
B	β	beta	[bet]	贝塔	磁通系数；角度；系数
Γ	γ	gamma	[gaːm]	伽马	电导系数（小写）
Δ	δ	delta	[delt]	德尔塔	变动；密度；屈光度
E	ε	epsilon	[ep'silon]	伊普西龙	对数之基数

续表

大写	小写	英文注音	国际音标注音	中文读音	意义
Z	ζ	zeta	[zat]	截塔	对数；系数；阻抗；相对黏度；原子序数
H	η	eta	[eit]	艾塔	磁滞系数；效率（小写）
Θ	θ	thet	[θit]	西塔	温度；相位角
I	ι	iot	[aiot]	约塔	微小、一点儿
K	κ	kappa	[kap]	卡帕	介质常数
Λ	λ	lambda	[lambd]	兰布达	波长（小写）；体积
M	μ	mu	[mju]	缪	磁导系数；微（千分之一）；放大系数（小写）
N	ν	nu	[nju]	纽	磁阻系数
Ξ	ξ	xi	[ksi]	克西	
O	o	omicron	[omikˋron]	奥密克戎	
Π	π	pi	[pai]	派	圆周率
P	ρ	rho	[rou]	肉	电阻系数（小写）
Σ	σ	sigma	[ˋsigma]	西格马	总和（大写）；表面密度；跨导（小写）
T	τ	tau	[tau]	套	时间常数
Υ	υ	upsilon	[jupˋsilon]	宇普西龙	位移
Φ	φ	phi	[fai]	佛爱	磁通；角
X	χ	chi	[phai]	西	
Ψ	ψ	psi	[psai]	普西	角速；介质电通量（静电力线）；角
Ω	ω	omega	[oˋmiga]	欧米伽	欧姆（大写）；角速（小写）；角

附录 N 读 法 实 例

$y=f(x)$ y is a function of x

$6\times 5=30$ six times (multiplied by) five equals (is equal to) thirty

$(x-y)(x+y)$ x minus y; x plus y

$\sqrt[5]{x^2}$ the fifth root of x square

y^{-10} y to the minus tenth (power)

$20:5=16:4$ the ratio of 20 to 5 equals the ration of 16 to 4 (20 is to 5 as 16 is to 4)

$e=1.6\times 10^{-19}$ e equals one point six multiplied by ten to minus nineteenth power

10^{-n} ten to the minus n

$\dfrac{1}{n^2}$ one over n square

$\dfrac{1}{1-nz^{-1}}$ one over one minus n times z reverse

$f(x)=ax^2+bx+c$ the function of x equals a times the square of x plus b times x plus c

$|a|=b$ the absolute value of a equals that of b

$\max f(x)$ the maximum value of $f(x)$

$\min f(x)$ the minimum value of $f(x)$

∞ infinity

$\lim\limits_{n\to\infty} S_n = \dfrac{1}{3}$ the limit of S_n as n gets arbi-

$\dfrac{x^5+A}{(x^2+B)^{\frac{2}{3}}}$ trarily large is one third

x to the fifth power plus A over (divided by) the quantity x squared plus B, to the two-thirds power

$(A+B)\ C$　the quantity A plus B times C
$A+B=C$　A plus B equals C
$A-B=C$　A minus B equals C
$A\times B=C$　A multiplied by B equals C
$A/B=C$　A divided by B equals C

附录 O　数学问题求解的一般表示

Solve the following system of equations
Solution：multiply equation (1) by (2) and get
Subtract equation (2) from equation (4), and get
Subtract equation (3) from equation (2), and get
Subtract equation (5) from equation (6), obtain x and y

附录 P　进　　制

radix 进制；基数
quanternary 四进制
binary 二进制
quinary 五进制
octal 八进制

senary 六进制
decimal 十进制
duodecimal 十二进制
hexadecimal 十六进制

参 考 文 献

[1] 赵永平 主编. 道路工程英语. 北京：人民交通出版社，1999
[2] 方旭明 编著. 新编专业外语. 成都：西南交通大学出版社，1997
[3] 李嘉 主编. 专业英语（公路、桥梁工程专业用）. 北京：人民交通出版社，1998
[4] 邓贤贵. 建筑工程英语（第2版）. 武汉：华中理工大学出版社，1997
[5] 张梦井，杜耀文 编著. 汉英科技翻译指南. 北京：航空工业出版社，1996
[6] 清华大学"英汉技术词典"编写组 编. 英汉技术词典. 北京：国防工业出版社，1978
[7] ASCE. Civil Engineering Magazine，2002~2004
[8] MacGinley T J. Steel Structures. New York：E & F N Spon Ltd，1981
[9] Bickel J O, Kuesel T R, King E H (Ed). Tunnel Engineering Handbook. Second Edition. New Tork：Chapman & Hall，1996
[10] 李亚东 编著. 土木工程专业英语. 成都：西南交通大学出版社，2005
[11] 葛耀君 编著. 土建英语. 上海：上海教育出版社，1996
[12] 朱梅心 编著. 港口及航道工程专业英语. 北京：人民交通出版社，1995
[13] 陆铁镛 主编. 测绘专业英语选读. 北京：测绘出版社，1991
[14] W F Chen (Editor in Chief). The Civil Engineering Handbook. Boca Raton：CRC Press，Inc.，1995
[15] 苏小卒 主编. 土木工程专业英语. 上海：同济大学出版社，2002
[16] 段兵廷 主编. 土木工程专业英语. 武汉：武汉理工大学出版社，2003
[17] 周开鑫 主编. 工程英语（土木类 教程）. 北京：人民交通出版社，2001
[18] Augustine J. Fredrich, Sons of Martha：Civil Engineering Readings in Modern Literature. New York：ASCE. 1989
[19] Jonathan Ricketts, M Loftin, Frederick Merritt. Standard handbook for civil engineers [M]. McGraw-Hill Professional，2003
[20] A T Papagiannakis, E A Masad. Pavement and Materials Design Manual [M]. Frederick Merritt John Wiley & Sons，2008
[21] The British Tunnelling Society and The Institution of Civil Engineers. Specification for tunneling third edition [M]. Thomas Telford Limited，2010
[22] Michael S, Mamlouk, John P, Zaniewski. Materials for Civil and Construction Engineers [M]. Prentice Hall，2006
[23] A R BIDDLE BSc, CEng, MICE. H-Pile Design Guide [M]. The Steel Construction Institute，2005
[24] Eoffrey Griffiths, Iclc Thom. Concrete Pavement Design Guidance Notes [M]. Taylor&Francis，2007
[25] Wai-Fah Chen, Lian Duan. Bridge Engineering Construction and Maintenance [M]. Taylor&Francis Group, LLC，2003
[26] Arnold Verruijt. Soil Mechanics [M]. Delft University of Technology，2001